Telecommunications Reform in the Asia–Pacific Region

Telecommunications Reform in the Asia–Pacific Region

Edited by

Allan Brown, Moazzem Hossain and Duc-Tho Nguyen

Griffith University, Nathan, AUSTRALIA

Edward Elgar

Cheltenham, UK • Northampton, MA, USA

Published by
Edward Elgar Publishing Limited
Glensanda House
Montpellier Parade
Cheltenham
Glos GL50 1UA
UK

Edward Elgar Publishing, Inc.
136 West Street
Suite 202
Northampton
Massachusetts 01060
USA

A catalogue record for this book
is available from the British Library

ISBN 1 84064 691 8

Printed and bound in Great Britain by MPG Books Ltd, Bodmin, Cornwall

Contents

Figures

Tables

Contributors

Risa Bhinekawati, former Executive Director, MASTEL, Jakarta, Indonesia.

Allan Brown, Associate Professor, Griffith Business School, Griffith University, Nathan, Australia.

Robert W. Crandall, Senior Fellow, Brookings Institution in Washington DC, USA.

Cosmo Graham, Professor and former Head, School of Law, University of Leicester, Leicester, UK.

Jerry Hausman, the John & Jennie S. MacDonald Professor of Economics, MIT, Cambridge, Massachusetts, USA.

Moazzem Hossain, Senior Lecturer, Griffith Business School, Griffith University, Nathan, Australia.

Sisira Jayasuriya, Director, Asian Studies, University of Melbourne, Melbourne, Australia.

Malathy Knight-John, Senior Researcher, Centre for Development Policy, Colombo, Sri Lanka.

Quoc-Phuong Le, Griffith Business School, Griffith University, Nathan, Australia.

Cassey Lee, Senior Lecturer, Department of Economics, University of Malaya, Kuala Lumpur, Malaysia.

Justin Malbon, Associate Professor, School of Law, Griffith University, Nathan, Australia.

Duc-Tho Nguyen, Professor, Griffith Business School, Griffith University, Nathan, Australia.

Rhoda U. Reyes, PhD Candidate, School of Film, Media and Cultural Studies, Griffith University, Nathan, Australia.

Ramonette B. Serafica, Associate Professor, Economics Department, De La Salle University, Manila, Philippines.

Nittaya Soonthonsiripong, Regulatory Analyst Manager, CP Orange, Bangkok, Thailand.

Koesmarihati Sugondo, former Secretary, MASTEL, Jakarta, Indonesia.

Nhat-Le Tran, Policy Department, Department General of Posts and Telecommunications, Hanoi, Vietnam.

Yanrui Wu, Senior Lecturer, Department of Economics, University of Western Australia, Nedlands, Australia.

Faizal Yahya, Lecturer, Department of Economics, National University of Singapore, Singapore.

Preface

Telecommunications industries around the world are experiencing fundamental changes. These include the advent of new (mainly digital) technologies, a rapid expansion in the range of telecommunications services available, increased competition by way of new players entering established as well as developing markets, interconnection of new entrants with the networks of incumbent carriers, and in many countries, full or partial privatisation of the previously government-owned monopoly carrier.

This wide-ranging process has produced an apparent paradox. Although the process has been generally (and fairly) characterised as a 'liberalisation' of the industry, in most countries where such changes have taken place, the liberalisation has been accompanied by, and interwoven with, a significant increase in the level, scope and complexity of government regulation. One possible explanation for this seeming contradiction is the need for regulators and operators alike to deal with rapidly evolving technologies, which in many cases have rendered previous boundaries untenable, and in many others have extended them far into unknown territories.

This book explores the twin themes of reform and regulation in the telecommunications sector of a dozen countries. It is divided into two parts: Part I examines the reform experience of three industrialised countries – the United States, the United Kingdom and Australia; and Part II provides case studies of developing countries in the Asia–Pacific region, namely, Indonesia, Malaysia, the Philippines, Thailand, Sri Lanka, Vietnam, and China and interactions between India and Singapore. A major aim is to draw lessons from the international experiences in telecommunications reform and, in particular, to identify the problems encountered and the mistakes made by developed countries, with a view to alerting developing countries to the potential pitfalls inherent in the entire process.

This aim was also the main motivation underlying an international workshop, on *Telecommunications Reform in the Asia–Pacific: Can Developing Countries Benefit from the Experiences of Developed Countries?*, which was held in Jakarta in November 2000. In a sense, the book arose directly from this workshop: the papers presented at the workshop on the United Kingdom, Australia, Indonesia, Thailand, Sri Lanka, Vietnam and China became earlier drafts of the corresponding chapters in this volume.

The current book went further than the workshop, however, in that new chapters were commissioned for the United States, Malaysia, the Philippines and India–Singapore in order to provide a more comprehensive coverage, with respect to both developed and regional countries.

While editors' names have been listed in alphabetical order, we share collective responsibility with regard to the planning and design of the workshop, and the editorial work in producing this book.

The Jakarta workshop was organised jointly by Griffith University (Brisbane, Australia) and the Telecommunications Association of Indonesia (MASTEL, Jakarta, Indonesia). We wish to record our sincere appreciation of the major role played by Mr Sukarno Abdulrachman, Chairman of MASTEL, and of the dedication and professionalism of MASTEL staff who worked tirelessly to ensure that the more than 100 delegates from different parts of Indonesia and the 15 international speakers enjoyed two memorable days of lively and productive discussion.

We would also like to acknowledge the financial support received from three of Griffith University's academic schools (Law, Economics and International Business) and two of its research centres (the Griffith Asia–Pacific Research Institute and the Australian Key Centre for Cultural and Media Policy). Their assistance was instrumental to the success of the Jakarta workshop. We wish to acknowledge, in particular, the personal support and encouragements from the then Heads of Schools – Professors John Dewar, Christine Smith and Bill Shepherd.

The financial support received by MASTEL from PT Tiara Titian Telekomunikasi, PT Telkomsel, PT Indosat and PT Margahayu Mapan Indocorpora in relation to the workshop in Jakarta is also highly appreciated.

Finally we wish to thank Robyn White for her perseverance and skills in supporting us and seeing the long and arduous editing process through to its conclusion.

Brisbane, October 2004

A.B., M.H. and T.N.

Summary

In recent years the US, the UK and Australia have extensively reformed their telecommunications industries. Reform in these countries has taken the form of (a) full or part privatisation of the incumbent monopoly carrier, (b) the introduction of new telecommunications carriers thus opening the industry to competition, and (c) the establishment of a detailed regulatory mechanism to promote competition in the industry.

This study provides a forum to bring together a number of stakeholders and academics in telecommunications policy from a number of developing and newly industrialising Asia–Pacific nations – namely, Indonesia, Malaysia, the Philippines, Thailand, Sri Lanka, Vietnam, China, India and Singapore. The cases of the US, the UK and Australia have been incorporated to share the knowledge and experience of telecommunications reform in these countries. The primary aim is to explore in relation to telecommunications reform: Can developing countries benefit from the experience of developed countries such as the US, the UK and Australia?

Specifically, the aims are:

- to draw lessons from cross-national experience in telecommunications reform and regulation;
- to examine the extent to which telecommunications policies can be transferred from mature developed economies to developing economies;
- to identify the problems encountered and mistakes made by the US, the UK and Australia in the reform of their telecommunications industries and so alert the developing countries of the potential pitfalls of the reform process; and
- to analyse the possible consequences of selected policy transfer in the new environment of converging markets, particularly in telecommunications, broadcasting and the Internet.

What are the benefits of such an exercise?

- to share knowledge and experience relative to telecommunications reform;
- to establish a network among telecommunications stakeholders in the Asia–Pacific region.

It is widely accepted that the development of Information Technology, of which telecommunications is a central part, is vital to the future social and economic well being of modern societies. The US, the UK and Australia have liberalised, privatised, and has regulated their telecommunications industries, and the developing Asia–Pacific nations have commenced or are about to commence their telecommunications reform process. Among other things, the present study provides a forum for representatives from the Asia–Pacific nations to learn from the experience of developed countries regarding the complex but important areas of liberalisation and regulation of telecommunications.

The specific benefits out of this study include:

- an understanding of the technical, legal and economic interrelationships, and the importance of transparency and accountability in the telecommunications reform process;
- an exposition of the issues of increased competition and market convergence in the telecommunications industries;
- the illustration of tensions between economic and non-economic objectives involved in telecommunications reform, especially employment issues and universal service obligations;
- the identification of the lessons from the telecommunications reform experiences of the US, the UK and Australia regarding measures that have been successful and mistakes that have been made; and
- an exploration of the application of the lessons from the developed countries to the process of telecommunications reform in the developing countries.

This book has emerged from the international workshop which the editors organised jointly on the above theme held in Jakarta in November 2000. Almost all the chapters were presented in the workshop by the respective authors except those on Malaysia, the Philippines, India–Singapore and the US. These chapters have been subsequently commissioned to make the regional and developed country coverage of the book more complete. Each chapter deals with the recent reform agenda of the respective nation and also presents the outcome of that reform. Part I demonstrates the experience of three developed nations: the US, the UK and Australia. Part II presents case studies from selected Asia–Pacific nations. A brief summary of each chapter has been presented in the rest of this introduction.

In Chapter 1, Crandall and Hausman present a clear picture on telecommunications reform in the post–1996 period in the US. In the prior 50 years the US telecommunications sector has been a regulated private monopoly, unlike the situation in other countries in the West where the telecommunications industry was a state-owned monopoly. Liberalisation in the US began in the 1970s with the FCC permitting competition in the market for interstate dedicated business connections. Major reform was brought to telecommunications when the US Congress passed the 1996 Telecommunications Act. This legislation opened local telecommunications markets to competition, seeking to complete the earlier market opening process in long-distance services and creating an economic environment leading to the 'deployment of advanced telecommunications and information technologies and services to all Americans'. This Act among other things has contributed to a heavy-handed regulation to promote competition in the network segment of the US industry. The wireless sector (cellular), however, thrived over the years without regulation. Crandall and Hausman conclude that 'detailed cost-based regulation of wholesale rates is not a satisfactory approach for stimulating competition in this network industry, particularly one in which technology is changing rapidly… Other countries, particularly those still requiring substantial investment in basic telecommunications infrastructure, would be well advised not to emulate the intensely regulatory approach adopted by the US'.

The telecommunications reform experience in the UK is presented in Chapter 2 by Graham and Hossain, analysing in particular the economic, legal and regulatory issues of reform. The UK reform began with the privatisation of the, then, British Telecommunications (BT) in 1984. Since this period the industry has experienced a constant process of reform in economic, legal and regulatory terms. For this purpose the British government established an independent regulatory agency called the Office of Telecommunications (Oftel) under the 1984 Telecommunications Act. Oftel is headed by a Director General, appointed by the government, with formal legal responsibility for the operation of Oftel's powers. Provision of telecommunications services was made subject to a licensing regime that is, the provision of telecommunications services became a criminal offence unless done so under a licence granted by the Department of Trade and Industry. Oftel's Director General, however, has other responsibilities in addition to those of licence modification and enforcement and these include promoting competition, ensuring quality of service and price control. Most commentators seem to regard the agency's performance since 1984 as successful. The price of telecommunications services has fallen, the quality of service has improved substantially, and customers now have a choice of suppliers and services. The regulatory system itself has developed in a notably open way, as far as Oftel is concerned, and it is much easier to follow the development of policy. It can safely be said that the UK system has to a large extent fulfilled the transparency and accountability objectives of the reform process.

Chapter 3 presents the major reform measures that have been taking place in the Australian telecommunications sector since the early 1990s. The main objective of reform is to introduce competition within an industry dominated by a formerly state-owned monopoly. To attain this objective the privatisation of Telstra began in 1997 with one-third privatised in the first tranche and another one-sixth in the second tranche in 1999. At present, 49.9 per cent of the shares in Telstra are owned by private shareholders and the remaining 50.1 per cent are held by the Australian government. Brown and Malbon argue that the government's objective to bring competition into the industry has a few intractable problems. The most important and fundamental of these is the structural imbalance in the industry arising from Telstra's ownership and control of the fixed local access network which gives the company substantial market power in call termination. This market power exists regardless of the number of industry players and Telstra's declining aggregate market share. This is because there is no alternative for call termination other than to use the network operator to which the called party has subscribed for local calls, usually Telstra. Therefore, the authors conclude that telecommunications reform in Australia has been suboptimal. It has left a structural imbalance that will continue seriously to impede competition and effective regulation of the industry. In other words, the government's policy goal of achieving substantial competition is unlikely to materialise with the given policy measures in place. It appears that the desired level of competition is unlikely to be attained in the foreseeable future.

In Chapter 4, Sugondo and Bhinekawati present the first country study, Indonesia, among the developing nations of the Asia–Pacific region. The government of Indonesia began reforming the telecommunications sector through an Act passed in 1989. During the early 1990s, the industry was subject to various changes and the government's economic and regulatory reform initiatives were considered broadly appropriate. In particular, the 1999 Telecommunications Act paved the way for a major restructuring according to a blueprint of telecommunications reform in Indonesia. This does not mean that the industry is not facing any challenges. Since the government has been acting as the major player in regulating the industry, and is also the owner of the principal incumbent companies, there is a conflict of interests in implementing the reform policies. It is observed that the implementation of reform policies has been very slow and in some cases the major players (government agencies) are hesitant towards reform, due to lack of confidence and lack of knowledge as to the benefits of liberalisation. The authors have concluded that this situation can be overcome by thorough informed discussions between the policymakers and the stakeholders, employees and members of the public. The authors also believe that to achieve reasonable success in regulating the telecommunications industry in Indonesia, 'the regulating authority should be empowered and transformed into a truly independent body'.

In Chapter 5 Lee presents the case of Malaysian telecommunications reform. Among the developing Asia–Pacific countries, Malaysia has in place a relatively efficient and effective telecommunications industry. Between 1985 and 2000, the country's telephone penetration rate rose by 540 per cent. The liberalisation and privatisation of the telecommunications sector in the 1980s ushered in an era of regulatory reforms and competition in the industry, beginning with the reforms of 1983 when the government allowed the private sector to compete with the Jabatan Telekom Malaysia (JTM) to supply equipment such as telephones and teleprinters. Since this period, in terms of bringing reforms into telecommunications, Malaysia has made genuine progress. The state-owned monopoly started relinquishing its special position in 1990 by selling 25 per cent of its equity to the public. Liberalisation in the post-privatisation era mainly took the form of entry into two major markets: fixed line and cellular services. Between 1993 and 1995, five additional licences to operate in the fixed line market were approved. The Communications and Multimedia Act of 1998 opened up the creation of an independent regulatory regime for the telecommunications sector. Under this Act a new regulatory authority, the Communications and Multimedia Commission (CMC) was established. The commission is comprised of five members. Its activities are classified into four key areas: economic regulation, consumer protection, technical regulation and social regulation. While a regulatory infrastructure is in place now, it is however, yet to perform and set market-based prices for various services. The Minister of Energy, Communications and Multimedia still has the power to decide on regulatory policies.

In Chapter 6, Reyes and Serafica present the reform experience in the Philippines which adopted a similar reform agenda to that of the US, since the country's telecommunications infrastructure during the twentieth century followed the features of the private monopoly system of the US. For more than 60 years the Philippines Long Distance Telephone Company (PLDT) had been operating as a private monopoly. This was broken when two more operators were allowed into the international long distance market in 1989. The liberalisation in telecommunications continued in the 1990s with

Presidential Executive Orders (EO 59 and EO 109) that introduced further competition. Finally, in 1995 the Philippines Congress passed the Republic Act (RA 7925) to further develop the industry and set out a regulatory framework for telecommunications. The liberalisation process since 1989 has made the telecommunications industry one of the most dynamic sectors of the Philippines economy. However, it has also experienced some major weaknesses over the years. There is a concern about regulating the industry effectively. The regulatory agency National Telecommunications Commission (NTC) lacks resources and executive powers. This makes the government's competition agenda ineffective since presently the dominant incumbent (PLDT) possesses an absolute market power. Establishment of an effective regulatory mechanism remains a major challenge.

The telecommunications reform experience in Thailand is presented by Soonthonsiripong in Chapter 7. Like most of the developing Asia–Pacific nations, Thailand also has a state monopoly in telecommunications. The government has undertaken a liberalisation programme to bring competition to the industry since late 1980s. In 1986, a programme called build-transfer-operate (BTO) was introduced to bring efficiency in running the telecommunications industry. Under the BTO scheme, private companies are granted concessions to build a telecommunications facility, but immediately upon completion of the project the participating companies are required to transfer the facility to public ownership. Upon receiving the ownership the government in turn allows the private companies to operate the respective facility for a fixed period. For example, Pacific Telesis has developed a paging service and has received ten years of concession to run it. By 1999 the number of projects with the BTO type arrangement reached 30. The BTO has lost its appeal in the liberalisation process over the years and subsequently in 2000, a State Enterprise Policy Committee was formed to convert the BTO concession agreements. The Committee is looking for an effective method to convert the concessionaires into private owners. However, it will take time to make this process of conversion work in Thailand. It is expected that the liberalisation and privatisation of the Thai telecommunications industry will be complete by 2006 in line with the country's agreement with the World Trade Organization (WTO).

Sri Lanka's experience in telecommunications reform is presented by Jayasuriya and Knight-John in Chapter 8. The telecommunications industry in Sri Lanka has been subject to various changes and reform since the 1980s, notably in two major areas: telecommunications structure and regulation. With restructuring of state-owned telecommunications, the deregulation of value-added telecommunications services and the entry of two new players into the fixed-access sector in 1996, Sri Lanka's telecommunications industry has been profoundly transformed, and has become one of the most open and competitive among developing countries in the Asia–Pacific region. The state-owned entity was partially privatised in 1997. Since then, it has been observed that the government is having second thoughts regarding pro-competitive reform agendas. The Telecommunications Act of 1996 put a regulatory agency Telecommunications Regulatory Commission (TRC) in place to foster competition and to accommodate the interest of a wide spectrum of stakeholders. However, in recent years it has been observed that the regulatory balance has tilted in favour of the dominant operator, Sri Lanka Telecom Ltd (SLTL). In Sri Lanka, the politicisation of the regulatory process made the regulatory authority more vulnerable to capture by a particular interest group and as a result becoming ineffective in conducting its functions. At this stage it is essential to manage the industry through true market forces instead of politically favoured competitors.

Chapter 9 presents the case of Vietnam, a relative newcomer in the field of telecommunications reform. Nguyen, Le and Tran state that Vietnam's teledensity was only approximately 0.14 fixed telephone lines per 100 people in the early 1990s. During that decade, however, dramatic improvements were made, with teledensity reaching 4 fixed lines per 100 people by 2000, and similar growth in geographical coverage and in other segments of the market were experienced, when consumer charges were reduced considerably. While all these outcomes have been encouraging, according to the authors the authorities have been confronted with a dilemma suggesting that, 'on the one hand, there has been an urgent need to rapidly expand and upgrade networks and services, which in turn implies a greater role for foreign and private domestic investors. On the other, the industry continues to be seen as vital to the country in non-economic as well as economic terms, such that the authorities have been determined to maintain a high level of effective control'.

In Chapter 10, Wu investigates telecommunications reform in China over the last two decades. During this period, the telecommunications industry has undergone dramatic changes in terms of regulatory environment, infrastructure development and provision of services. As a result of these changes, the industry has expanded significantly and China now has the world's second largest number of telephone lines after the US. Prior to 1998, several government departments were involved in both operation and policymaking in the sector. The Ministry of Posts and Telecommunications (MPT) was the major player and owned, via China Telecommunications Company Limited (China Telecom), the bulk of China's fixed lines. Separate 'private' networks were also owned and operated by other government departments, namely the Ministry of Railways (MOR), Ministry of Electric Power (MEP), Ministry of Defence (MOD) via the People's Liberation Army, and the Ministry of Petroleum (MOP). A major reform in the industry took place in 1994 with the establishment of a second national carrier, United Telecommunications Corporation (Unicom), which became a direct challenge to the monopoly held by China Telecom. Changes to the organisational structure for telecommunications were initiated in 1998. The Ministry of Posts and Telecommunications, Ministry of Electronic Industry and the State Radio Regulation Commission were merged to form the present Ministry of Information Industry (MII). The aims of this change were clearly to consolidate the government's regulatory role and create an environment for effective competition. Furthermore, initiatives have been taken to break up China Telecom's four service divisions into separate entities; four separate operating companies are now responsible for fixed line, mobile, paging and satellite communication services. In recent times, China has introduced new policy initiatives to deregulate the industry, particularly the value-added part. As far as the fixed line services is concerned, deregulation is far from complete. The first Telecommunications Act was introduced as recently as September 2000, and enforceability is yet to be tested. However, it is encouraging to see the government taking steps towards freeing the telecommunications market to foreign investors starting in the late 2001 after China's entry to the WTO.

Faizal Yahya contributes the final chapter which includes an investigation into investment initiatives of Singapore-based companies in the rapidly expanding telecommunications market in India. In doing so, Yahya analyses, first the continuous reform process taking place in India's telecommunications sector and subsequently highlights the opportunities the reform creates for attracting further foreign direct investment. Next, a brief review of Singapore's national telecommunications carrier (SingTel) is provided, particularly looking at its investment ventures in telecommunications all over the world. SingTel's investment in the Asia–Pacific region, however, receives major attention in the study. Finally, this company's joint venture initiatives with its Indian private counterparts are investigated. So far, SingTel has pumped over US$400 million into India. In the future, it has a plan to provide further funds to expand the Indian telecommunications infrastructure. It is very interesting to see a foreign government owned telecommunications company working hand-in-hand with privately owned counterparts in the region. Furthermore, the chapter demonstrates that, due to its comparative advantage, Singapore has the resources and will to invest in other Asian economies and to exploit the synergy between them for their mutual benefit. This is an eye opener for other Asia–Pacific economies in relation to attracting foreign direct investment in telecommunications from the prosperous economies of the region.

PART I

DEVELOPED NATIONS

1. United States: Lessons for the new millennium

Robert W. Crandall and Jerry Hausman

For more than fifty years the United States telecommunications sector was a regulated *private* monopoly, dominated by the American Telephone and Telegraph Company (AT&T). This fact alone distinguishes the US experience from that of most other countries, which typically had *state-owned* telecommunications monopolies. During most of the period from 1934 to 1984 the US Federal Communications Commission (FCC) and a variety of state authorities controlled the relative prices of telephone services and barred the entry of competitors. As the pressure for liberalisation and deregulation spread from other sectors of the US economy the support for this government-contrived monopoly weakened. A major antitrust suit led to the dissolution of the country's national carrier, AT&T, in 1984, but entry into the provision of local telephone services did not occur until the passage of the 1996 Telecommunications Act.

1. THE PATH TO LIBERALISATION AND COMPETITION

In the 1970s the first breath of liberalisation swept over the sector as the FCC began to permit limited competition in the market for interstate dedicated business connections. This competition spread eventually to all long-distance services, against the FCC's wishes. At the same time, the FCC won a battle with state regulators to open the market for terminal equipment (telephone handsets, answering machines, modems, etc.) to competition. The states had challenged the FCC's right to invalidate state tariffs that required the use of telephone-company supplied terminal equipment. Presumably, the state regulators saw competition in terminal equipment as a threat to their attempt to maintain high rates for company-provided business equipment in return for low local residential rates.

The path to increased competition in long-distance services and terminal equipment was not without its problems. When AT&T used its control of local facilities to frustrate competition in long-distance services and terminal equipment, a major antitrust suit was filed by the federal government resulting in the 1984 divestiture from AT&T of its local operating companies. The antitrust decree imposed a quarantine on the divorced local Bell operating companies that prevented them from offering long-distance services in competition with the national long-distance carriers – AT&T, MCI and Sprint – or from manufacturing telephone equipment such as handsets, modems and switches.

After the 1984 breakup the terminal equipment market became very competitive. However, competition in local telecommunications grew very slowly because state regulators limited entry and controlled local rates, and long-distance rates remained substantially above their competitive levels. For twelve years the AT&T trial court wrestled with a large number of difficult issues in implementing the decree, and the regional Bell companies chafed at their continued exclusion from long-distance services. The long-distance carriers were equally concerned about the slow progress towards competition in local markets, a problem beyond the reach of the AT&T decree.

As a result of these concerns, the US Congress was finally prodded to reform the entire telecommunications regulatory structure through the passage of the 1996 Telecommunications Act. This legislation opens local telecommunications markets to competition, seeks to complete the earlier market opening process in long-distance services (including freeing the Bell operating companies from their quarantine), and creates an economic environment that will lead to the 'deployment of advanced telecommunications and information technologies and services to all Americans'.

The 1996 Act invites regulators to 'forbear' from regulation of new services whenever such forbearance is in the 'public interest', but unlike earlier exercises in deregulating network industries, the Act provides a massive increase in the scope of regulation and an unfortunate return to cost-based regulation.[1] Moreover, the Act not only retains a system of regulatory cross-subsidies erected by the states and the FCC over several decades – popularly described as 'universal service' policies – but it also extends them to new services and to new recipients. Furthermore, the Act requires that the federal share of these subsidies is to be funded by an explicit charge or tax levied on telecommunications company service revenues.

Despite the early legal and regulatory uncertainties that have plagued the implementation of the Act, local competition has begun, albeit slowly. However, a simpler framework, under which local competition might have developed even more quickly and the restrictions on Bell company entry into long-distance services might have been lifted sooner, would have promoted competition without risking deleterious effects on the investment in new services by incumbent local carriers. The enormous economic gains that are potentially available from the development of these new services and from the movement of the prices of traditional telecommunications services to competitive levels are being squandered in the belief that detailed regulation is the best route to competition. A much less regulatory regime, stressing open entry and a much simpler regime of reciprocal compensation for interconnection, would have been far more effective in promoting entry and investment.

2. MARKET STRUCTURE AND THE NEED FOR FURTHER LIBERALISATION

The US was once the leader in opening its telecommunications markets to competition. The FCC allowed limited entry into interstate long-distance services in 1969–71, although entrants were allowed to offer only dedicated, private lines to business customers. In the mid-1970s MCI audaciously entered ordinary, switched long-distance markets without FCC permission, and fended off the FCC's attempts to block its entry in federal courts. As a result, interstate long-distance competition began to develop in the late 1970s.

The states regulate intrastate services, including the regular flat-rate local service that most residences and small businesses use to connect to the network. For the most part, until 1996, state regulators discouraged or blocked entry into these local services and even into intrastate long distance so as to maintain a politically-crafted pricing policy of subsidising residential local service from long distance and business service charges.

Table 1.1 US telecommunications sector, 1999

Service	Revenues ($ billion)
Local	78.6
Long distance	93.3
Wireless	43.8
Total	215.8
Total revenues as percentage of GDP	2.3%

Note: Revenues are end-user revenues only and do not include revenues from wholesale services sold to other carriers.

Source: Lande (2000)

The US telecommunications sector generates more than $200 billion in annual revenues, an amount representing more than 2 per cent of gross domestic product (GDP) (see Table 1.1). Long-distance services are still nearly half of telecommunications revenues, and wireless (cellular) services have risen to more than 20 per cent.

2.1 The AT&T Case

US competition policy is largely responsible for the current structure of the US telecommunications sector. In 1974 the Justice Department filed a Sherman Act suit against AT&T alleging that AT&T had used its control of access to local customers to monopolise long-distance services and terminal equipment. This suit was largely the result of disputes that arose over entrants' attempts to interconnect with AT&T's local companies to complete their long-distance calls or to attach their terminal equipment. Had the FCC pursued a deliberate policy of allowing entry and requiring that AT&T provide equal access to its network for all competitors, these disputes might have been avoided and the antitrust action rendered unnecessary.

The AT&T litigation resulted in a settlement in 1982 that required AT&T to divest its local operating companies into seven separate regional Bell operating companies (RBOCs) two years later. The RBOCs were barred by the resulting decree from offering long-distance service outside their local areas, from manufacturing equipment, and from offering information services. The information services ban was overturned by an appellate court in 1991, but the other restrictions remained in place until 1996. The decree required the divested local Bell companies to reconfigure their network switching systems so as to interconnect all long-distance carriers on equal terms. The result was a substantial acceleration in the rate of expansion of the long-distance entrants and a corresponding decline in AT&T's share of long-distance revenues.

2.2 The Long-distance Market

The US long-distance market was dominated by AT&T prior to the 1984 antitrust divestiture. AT&T accounted for more than 85 per cent of all interstate and intrastate long-distance revenues before it was split into a long-distance company and seven regional operating companies. Since that time long-distance revenues have grown from $51 billion to $93 billion, and competition has increased markedly. The Hirschman–Herfindahl index of concentration, based on long-distance carrier revenues, has declined steadily from more than 8000 in 1984 to 2580 in 1999. However, it is still far above the Department of Justice's merger guidelines standard of 1800 for a highly-concentrated industry, equivalent to a market with about six equal-sized competitors (FCC 2000a). (A level of 2580 is roughly equivalent in competitive effect to a market with just four equal-sized competitors.) The three largest companies, AT&T, MCI/WorldCom and Sprint account for 76 per cent of long-distance carrier revenues, and there are at least 35 others with annual revenues of more than $100 million per year. During 2000, MCI/WorldCom and Sprint proposed to merge. The merger was blocked by US competition authorities who decided it would lead to higher long-distance prices and reduced competition in other areas.

In the US, the local companies offer intrastate long-distance services, and these companies now account for approximately 10 per cent of all long-distance service revenues, substantially less than before the 1996 Act. For decades state regulators had blocked competition in intrastate markets, but the 1996 Act required that even these markets be opened to competition by 1999.

2.3 Local Competition

While the long-distance markets had at least a moderate degree of competition by 1996, competition in the local-services market had only just begun at that time. Entry into local telecommunications had been allowed in a limited range of services, generally to connect large businesses to (interstate) long-distance carriers, including long-distance companies. As late as early 1996, only six states allowed entrants to serve the dispersed residential and small business subscribers. By limiting entry, state regulators were able to erect a rate structure for local connections that favoured residential subscribers, particularly those in rural areas or the outskirts of urban areas. Business lines, intrastate long distance, interstate access (for long-distance carriers) and advanced services have been generally priced substantially above long-run incremental cost so that residential lines can be priced below long-run incremental cost, particularly in rural areas.

Beginning in the 1980s, a number of new carriers began to build high-capacity fibre-optic ring networks in most major business districts in the US. These companies, known as competitive access providers (CAPs), generally marketed only to mid-size and large businesses because of state regulation, which kept residential rates generally below cost and generally discouraged entry.[2] After the passage of the 1996 Act a large number of other new local competitors emerged, building some facilities but relying heavily on resale of the incumbent local carriers' services. These new competitors are now known as Competitive Local Exchange Carriers (CLECs).

2.4 Potential Gains from Further Increases in Competition

US regulated telecommunications rates result in very large economic losses, due in large part to regulators' desires to keep the rates for local residential service very low, particularly in rural areas, and to cover the costs of these services from over-priced long distance and business services. Figure 1.1 shows the average 1996 US flat residential rate and two alternative estimates of local exchange carrier long-run (average) incremental costs (AIC), each arrayed by density of lines.[3] (We critique this model later in the chapter. For the present we simply use it to provide a lower bound on local exchange service costs.) It can be seen from the figure that clearly rates depart substantially from estimated AIC, regardless of the model used to estimate costs. In most states rates *rise* with increasing density, but the cost per line obviously declines. The resulting deficit of $5.5 billion to $16.2 billion per year is recovered in large part from (above-cost) long-distance charges.[4] Were residential rates moved to the estimated AIC per line and long-distance charges reduced by the amount of revenues raised, there would be an improvement in annual economic welfare of $2.5 billion to $7.0 billion per year. The potential welfare gains from full Ramsey pricing would obviously be greater. These losses are only from the distortions in *residential* rates. The economic costs of overpriced business services may even be higher. Businesses account for approximately 60 per cent of all long-distance calling and are faced with regulated prices of advanced services that are generally substantially above cost.

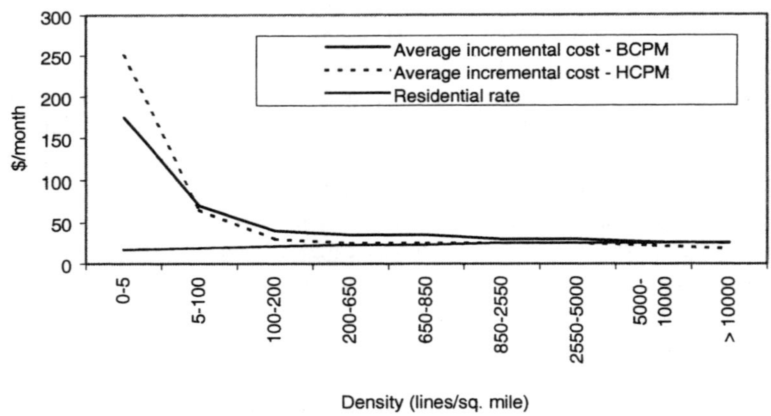

Source: Crandall and Waverman (2000)

Figure 1.1 Average US local residential rates vs average incremental cost

The regulation of telecommunications has also cost the US dearly by delaying new services. The electronics revolution has generated enormous opportunities for the development of new telecommunications services. Unfortunately these services must run a regulatory gauntlet before commercial introduction due to regulators' desire to protect existing service providers or certain consumers from the effects of competition and the loss of 'universal service' subsidies for traditional services. Cellular telephony, voice mail, video service delivery over telephone networks, and new satellite services have been delayed and (in the case of telephone-company delivery of video) even thwarted altogether by state and federal regulators.

Cellular telephony could have been licensed by the FCC in the early 1970s, but the first system did not begin operation until 1983 because of the FCC's delay in licensing the requisite frequencies. This delay in licensing cellular has been estimated by Hausman to have cost consumers \$33.5 billion (1994 dollars) per year (Hausman 1997). Using a similar methodology Hausman estimated that regulatory delay in allowing AT&T or its RBOC progeny to offer voice messaging services between 1981 and 1988 was \$1.3 billion per year (1994 dollars) (Hausman 1997). There are no estimates of the cost of regulatory delays in offering direct broadcast satellite services, telephone-company video services, or new broadband services over local telephone company lines, but they are likely to have been substantial.

3. THE 1996 TELECOMMUNICATIONS ACT

The US telecommunications sector operated for 12 years under the balkanised structure required by the AT&T decree. Widespread dissatisfaction with the lack of progress towards competition and the disputes that arose from enforcing the decree led Congress to pass the 1996 Telecommunications Act, which was signed into law in February 1996. The objective of this legislation was to open all telecommunications markets to competition, but the Act is far from an example of deregulation. Indeed, the Act erects a variety of new regulatory requirements that are supposed to guide, facilitate or otherwise stimulate competition. The most important of these provisions involves opening the networks of existing 'incumbent' local exchange companies (ILECs) to rivals so that the entrants can interconnect with them in any of a variety of ways. The incumbent companies are required to 'unbundle' their networks into a set of components or functions and to offer them to their prospective rivals at cost-based prices.[5] In addition, entrants are to be afforded the opportunity to resell the incumbents' retail services by obtaining these services from the incumbents at their retail prices less the avoided cost of retailing the service. Finally, incumbents must exchange local traffic with entrants on equal terms. All of these 'interconnection' terms are to be negotiated or, failing negotiation, are to be determined in arbitration proceedings conducted under the auspices of state regulators.

The Act's requirements for interconnection may be viewed as the 'stick' over the incumbent local exchange carriers' heads, requiring them to assist in opening their erstwhile monopoly markets to competition.[6] In addition, the Act provides a 'carrot' for the regional Bell local companies, allowing them to enter the long-distance market (from which they were banished by the 1984 AT&T antitrust decree) if they meet a detailed 14-point 'checklist' of market opening requirements. Few carrots have been harvested under this provision, however, for reasons we detail below.

The 1996 Act provides much more than a prescription for regulated competition in telecommunications: it makes major changes in universal service policy; mandates new subsidies for schools, libraries and rural health facilities; substantially deregulates cable television rates; liberalises broadcast ownership rules; and even regulates entry into the provision of alarm services. The universal service policies are to be supported by fees (taxes) levied on all telecommunications services, and are to be portable so that new entrants can receive the same payments as incumbents for offering services in areas where rates are below cost. In this chapter, however, we focus only on the provisions directly related to opening traditional telecommunications services markets to competition.[7]

3.1 Unbundling and Wholesale Rates

Competitive telecommunications networks can only survive if they interconnect with one another. Otherwise each subscriber would have to use separate terminal equipment for each of the competitive networks, an obvious inconvenience. No subscriber would want to have four or five separate telephone lines, handsets, and modems in his or her home or office. Interconnection however can take a variety of forms.

The simplest form of interconnection is simply the exchange of traffic. Indeed, national and international carriers, internet service providers, and adjacent local telephone carriers already exchange traffic at a variety of switching centres with apparent ease. The cost of such exchanges for voice-grade traffic is now measured in the tenths of a cent per minute, although international exchanges often take place at prices of 50 cents or more per minute because of a lack of competition in the telecommunications sectors of many countries. These rates are falling dramatically with the onset of liberalisation in many countries and the prodding of the FCC.

The 1996 Act provides a prescription for interconnection that is very different from the simple exchange of traffic. Using somewhat vague language, the Act requires local telecommunications carriers to interconnect at 'any technically feasible point' and to provide access to their 'unbundled network elements' (UNEs) at just and reasonable prices that are 'determined without reference to a rate of return or other rate-based proceeding'. An unbundled network element is a network component or function that a competitor may lease in order to deliver its services. Indeed, under the FCC's rules implemented in August 1996, a competitor could lease everything from its incumbent rival, thereby obviating the need to invest in any network facilities whatsoever.

The FCC further required that these elements be leased to entrants at 'forward-looking' costs, using an approach called 'total element long-run incremental cost' or TELRIC. Although these wholesale, carrier-to-carrier rates are to be part of the interconnection agreement negotiated between each entrant and each incumbent on a state-by-state basis, the rates are supposed to be guided by the FCC's forward-looking cost model or a similar model developed for each state.

These unbundling and wholesale pricing requirements have generated enormous controversy, resulting in numerous court challenges and a recent Supreme Court decision reversing a part of the FCC's rules. Specifically, the Court questioned the broad, sweeping requirement that every network element be provided to competitors.[8] However, the Supreme Court upheld the Commission's right to prescribe the cost standard for pricing these network elements, but the Court was not asked to address the question of how such costs could possibly be estimated by regulators. The FCC has subsequently relaxed the unbundling requirements only modestly while extending them in new directions, such as the requirement that incumbent companies actually share customer access lines with entrants rather than simply lease them.

3.2 Resale

In addition to unbundling, the 1996 Act requires local carriers to allow entrants to resell their services. Such resale simply transfers the marketing and billing functions from the existing local carrier to the new (reselling) entrant. The technical details of the service are unchanged; only the identity of the seller changes. Incumbent local carriers must offer all of their retail services to their competitors for resale, charging them their full retail price less a discount for the 'avoided costs' of marketing and billing the service. These discounts have typically been established through state-by-state negotiations and arbitration proceedings in the range of 15 to 25 per cent.

The resale requirement has not been as controversial as the unbundling requirements because resale does not provide the entrant with as much freedom to offer a bundle of new services with the assistance of the incumbent's facilities. However, substantial debates have arisen over the avoided-cost discounts needed to make such resale profitable. The FCC even forced SB Communications Inc. to increase its resale discounts as a condition for approving the recent SBC acquisition of Ameritech. The increased discounts were not based on the results of any economic analysis, but were a required payment to allow SBC to proceed with the acquisition.

The 1996 Act is noticeably silent on retail telecommunications prices except for mandating that explicit rural subsidies be sufficient to keep local rates in high-cost rural areas at levels comparable with urban rates. Incumbent carriers' intrastate services are still regulated by state commissions, and most of these commissions continue to administer a distorted rate structure (recall Figure 1.1). Intrastate long-distance rates and access charges, as well as interstate access charges, continue to cover the deficit between local per-line costs and local rates. While the 1996 Act prescribes cost-based wholesale rates, it does not require the state commissions to move retail rates towards cost. Indeed, the Commission has increased the distortions between retail rates and costs by increasing access charges to fund the internet subsidies to schools and libraries. As expected, these increased access charges have been passed on to consumers by the long-distance carriers.

3.3 New Services and Forbearance from Regulation

The 1996 Act has a provision (Section 706) that would permit a local carrier to escape regulation altogether if the FCC makes an affirmative decision to 'forbear' from regulating an 'advanced' telecommunications service. This provides the incumbent carrier with at least a theoretical possibility of building new, advanced facilities which may offer services without the burden of wholesale or retail regulation. Unfortunately, the FCC's only proposal for such forbearance requires that incumbents use structurally separate subsidiaries to deliver the unregulated, advanced service.

3.4 Long-distance Entry by the Bell Companies

The Act explicitly frees the Bell operating companies from the restrictions on long-distance entry imposed by the 1984 AT&T consent decree, but only after they satisfy a detailed checklist of market opening requirements spelled out in Section 271. In practice, this means that each company must pass through a series of three regulatory proceedings for each state – one at the state level, one at the US Department of Justice, and one at the FCC. The Section 271 requirements are generally labelled 'interconnection' requirements, but they are in essence a large number of requirements for non-discriminatory access to services, facilities, and unbundled elements. As of this writing, only two Bell companies – Bell Atlantic and SBC – have been able to satisfy the three regulators that it has achieved such non-discrimination along each of the Act's prescribed dimensions, but only for a total of six of the country's 50 states. For the most part, Section 271 approval has been stalled by entrants' complaints that the Bell companies have failed to install 'operating and support' systems that would allow for instantaneous transfer of customers from the Bells to the entrants at a very high transfer rate.

4. CRITIQUE OF THE 1996 ACT

While we cannot provide analysis of the problems in the design of the 1996 Act in this chapter, we list the more important failings in the hope that other countries might learn from the US mistakes. We note, moreover, that the Telecommunications Act differs fundamentally from earlier legislation that was passed to liberalise trucking, airlines, air cargo and railroads in that it expands the scope of regulation in the name of 'deregulation'. Wholesale rates and universal service subsidies are now to be determined by cost models. The Act opens all telecommunications markets to competition, even the erstwhile protected local markets, but it establishes a framework for entrant–incumbent cooperation in facilitating entry that has not been prescribed for any other of the recently deregulated sectors of the economy. The result has been five years of protracted regulatory disputes and litigation that show little sign of abating.

4.1 Unbundling

The requirement that incumbents unbundle their networks and lease the pieces ('elements') to entrants is rooted in the fact that local telecommunications networks are expensive to build, and certain parts of them may be 'essential facilities' or monopoly bottlenecks. But this provision was enacted without any guidance from historical experience. While it may be uneconomic for entrants to build parallel connections to subscribers' facilities, there is no empirical evidence on this matter, and there is certainly no reason to believe that *all* local network facilities are essential bottlenecks. The FCC's first order, requiring the unbundling of *everything* was too sweeping, but it has not recanted despite considerable evidence that entrants can and do build their own facilities. Competitors certainly can install their own switches, inter-office transport facilities and network intelligence. Wireless and long-distance companies do so routinely; why should local (wireline) competitors be any different?

A more troubling aspect of the unbundling requirements, however, is their imposition in a sector in which technology is changing extremely rapidly. Today's network is not static; it is steadily evolving into new, more advanced networks, such as those designed to deliver high-speed internet services. If an incumbent local network operator leases part of its current network to its rivals, will it be able to adopt new technology without these rivals' (entrants') assent? Surely, one would not want these new competitors using the regulatory process simply to delay or frustrate their rivals' attempts to innovate. Furthermore, extensive unbundling at below-cost prices reduces the investment incentives for entrants and incumbents alike. Given the uncertainty regarding technology the current regulatory approach is likely to slow down the adoption of new technology in the telephone network.

4.2 Regulating Wholesale Rates

Cost-based regulation has been in disfavour for a number of reasons that we need not review here. Although not the primary problem, an important problem is the difficulty that regulators have in determining the costs of providing the regulated service efficiently. How can federal or state regulators possibly know if arbitrated rates reflect the forward- or even backward-looking cost of each network element provided by an incumbent local carrier? Regulators are generally the last to know the level of costs, particularly in a dynamic industry such as telecommunications with its abundance of joint and common costs. Indeed, the FCC has spent most of the past five years building a detailed engineering model of forward-looking costs of an 'optimised', but probably outdated network to use in guiding its universal service policy. But this model is extremely controversial and is not generally employed. As we shall see, the federal–state regulatory process has already generated wildly different rates for wholesale elements across the states.

The forward-looking approach to estimating costs, even if it could be executed with accuracy, does not take account of the sunk and irreversible costs of network investment and therefore leads to below-cost unbundled element rates as Hausman has demonstrated in a number of papers. Instead, the FCC approach assumes, incorrectly, that all network investment costs are only fixed costs, not sunk costs, which is sharply at odds with economic reality. This approach gives a 'free option' to the entrants, and decreases the economic incentive of existing carriers to invest in network improvements, because it limits the economic returns to new investment. In addition, the FCC's forward-looking cost approach (TELRIC) provides an economic disincentive for investment by even new entrants into local telephone competition because it provides them with a subsidised price for the use of unbundled elements.[9]

The FCC's approach uses a 'contestable' market model for the regulation of unbundled elements, which assumes no sunk costs.[10] This approach does not recognise the interaction of sunk investments and the considerable technological and economic uncertainty that exists in telecommunications. The result is a set of prices for the unbundled elements that are 'too low', that is, below economically efficient levels, tilting the 'rent' or 'invest' decision of new entrants toward the rent decision, rather than the invest decision.

Suppose a competitor wants to lease the unbundled elements associated with this investment. The incumbent could offer the new competitor a contract for the economic life of the investment, say 10 years for investment in the local subscriber line or 'loop'. The price of the unbundled element would be the total investment cost plus the operating costs each year for the unbundled element. If demand did not materialise or if prices fell, the new entrant would bear the economic risk of this outcome. (The contract, or regulation, could allow the new entrant to sell the use of the unbundled element to another firm if it decided to exit the business.) However, the FCC's approach to setting wholesale rates at forward-looking cost typically allows the new entrant to lease the unbundled element on a month-by-month basis. If demand does not materialise or if prices fall, the incumbent has to bear the risk for the business case of the new competitor. Thus, the incumbent has been required by regulation to give a *free option* to the new entrant, namely, the right but not the obligation to purchase the use of the unbundled elements. The monthly price of the unbundled element should be significantly higher than the price of the element based on a 10-year life because the monthly price must reflect the risk inherent in the sunk investments or, equivalently, the value of the option given to the new entrant. Regulators to date have not incorporated the value of the option, which arises from the sunk cost nature of much telecommunications investment, into their price setting.

4.3 New Services

The 1996 Act allows the FCC to forbear from regulation of advanced telecommunications services if such forbearance advances the 'deployment on a reasonable and timely basis' of these services. However, the FCC's approach to this problem is entangled with the controversies surrounding network unbundling and restrictions on Bell-company provision of long-distance services between the local access and transport areas (LATAs) established to quarantine the Bell companies after the 1984 AT&T divestiture. To date, the FCC has agreed to forbear from regulating advanced services provided by the local carriers, but only if these carriers provide such services through separate subsidiaries. It has relaxed the long-distance restrictions on Bell companies only in those cases where such restrictions are demonstrably impeding the delivery of advanced services to rural areas. Thus Bell companies cannot build advanced networks that cross LATA boundaries. In addition, the FCC recently ruled that local carriers must share their access lines with competitive carriers wishing to offer only advanced, high-speed services. (Under this 'line sharing' regime the entrant leases only the higher frequencies on the line, leaving the lower frequencies to the incumbent for the delivery of traditional voice services.) The result is a set of polices that continues to regulate incumbents' advanced services and the leasing of their facilities to entrants offering similar services, but leaves the entrants' services unregulated. These regulations, promulgated largely to accelerate entry, are likely to reduce the incentives for the incumbent local carriers to invest in new advanced data networks that would provide faster internet access to 'all Americans' as called for by the 1996 Act.

5. THE US TELECOMMUNICATIONS SECTOR AFTER THE 1996 ACT

The 1996 Act was designed to open all telecommunications markets to competition, even the local and intrastate markets that had been protected by state regulators for decades. Since 1996 the large, incumbent Bell companies have lost some market share to the new local entrants, but in the growing economy of the 1990s this loss of lines to competitors was hardly noticed. The large Bell companies' access line growth was indistinguishable from the national growth rate in the late 1990s. The incumbents did not begin to suffer net losses of customers until early 2000.

5.1 Local Competition

By the end of 2000 the competitive local carriers had garnered about 9 per cent of the country's local telephone access lines (see Table 1.2). New entrants such as Time Warner, McLeod, RCN, WinStar and Allegiance have been expanding rapidly, but they do not yet account for a large share of the country's access lines. In addition, the older CAPs, such as Teleport and MFS, continue to expand. However, slightly more than four years after the passage of the 1996 Act, all of these competitors had succeeded in capturing only about 3 per cent of the country's telephone access lines on their own facilities, and another 5.5 per cent by either reselling the established carriers' services or leasing some of the established carriers' facilities.

Table 1.2 Number of access lines by source of facilities, December 2000 (in millions)

Carriers	Through own networks	Through resale or lease of incumbent facilities	Total
New competitors	10.7	5.7	16.4
Incumbents (largely Bell companies)	177.4	–	177.4

Source: FCC (2001)

Because the competitors have generally targeted business customers, the new competitors' share of local telecommunications *revenues* is much greater than their share of lines. Indeed, many of the new competitors bypass the residential subscriber completely because residential customers are so geographically dispersed and often are purchasing their local service from the regulated incumbent carriers at rates that state regulators have held below cost. As shown below, business rates are generally set much higher by state regulators in order to generate the revenues to subsidise residential connections. Moreover, businesses are likely to be much more concentrated geographically, making them attractive targets for the new competitors.

For several years the entrants have been waiting for an even more attractive regulatory option, namely, the entire bundle of network facilities in a (reconstituted) package of unbundled network facilities at regulated prices that are well below the resale rate. The pursuit of this goal led to a protracted period of legal and regulatory disputes over the right of entrants to demand the entire 'UNE platform', that is, the full complement of the incumbents' facilities at deeply discounted (TELRIC) rates. As part of its negotiations with the FCC to gain approval of the merger with NYNEX Mobil [sic] Communications Company, Verizon agreed to offer the entire wholesale platform in New York, resulting in a sharp increase in the number of leased local lines there. Competitors had amassed 20 per cent of subscriber lines in New York by the end of 2000, while they had only obtained an average of 7.6 per cent in the rest of the country (FCC 2001, Table 6).

5.2 Wholesale Rates

The most contentious issues in the implementation of the 1996 Act have been the FCC's decision to assume jurisdiction for establishing the rules for wholesale service and its prescription of forward-looking (TELRIC) prices for unbundled elements. Much time and energy have been devoted to regulatory and legal challenges to these decisions. The result has been a confusing array of state arbitrations of interconnection agreements that are still hotly contested. Indeed, there is often no agreement among the parties as to the current level of rates for unbundled elements or wholesale discounts across the states.

The data on wholesale rates shown in Table 1.3 have been assembled from a variety of industry sources. These rates are for business lines, and the unbundled loop rate is for the most dense, urbanised market in the state if the unbundled rates vary by population density. We focus on these business rates in the dense markets because the urban business subscriber is the most important target for the new competitors. Despite the fact that rates are apparently subject to a most-favoured-entrant or 'pick-and-choose' option, reported rates often differ substantially across entrants. We therefore show the rate for the largest entrants because they are likely to be the most important sources of competition.[11] Note the large variance in wholesale rates despite the fact that the technology employed by the incumbents does not differ much across states, particularly in the largest urban areas. For example, the unbundled loop rates range from less than \$5 per month to more than \$30. That is, a new entrant can lease a line in downtown Chicago for less than \$5 per month, but must pay more than \$25 per month for a similar line in downtown Denver. Obviously no 'cost model', however imprecise, guides these results.

Table 1.3 Distribution of RBOC unbundled business loop rates (UNEs) and wholesale discounts for business service, lower 48 states, 1998

Unbundled business loop rate: most dense area (\$ per month)	Number of states	Share of access lines (%)
Less than 10	6	20.8
10 – 14.99	16	48.0
15 – 19.99	16	22.2
20 – 24.99	6	5.5
25 or more	4	3.6
Wholesale discount for business service (%)		
Less than 15	9	6.2
15 – 17.49	11	35.4
17.50 – 19.99	14	32.0
20 – 22.49	12	22.8
22.5 or more	2	3.7

Sources: Janney Montgomery Scott Inc. (1997); industry sources

Similarly, wholesale discounts, which are supposed to reflect the avoided costs of marketing and billing, range from 12 per cent to 26 per cent of the business rate. This wide range might be understandable if it were inversely correlated with retail rates since the required discount would be lower for states with high rates. However, the resale discount is directly correlated with retail rates, but the correlation is not statistically significant. Obviously, the wholesale rates that have resulted from this system of regulated negotiation do not reflect any systematic measure of cost, forward- or backward-looking. Rather, the resulting prices are the outcome of a political process wrapped in the guise of efficient regulation.

5.3 Retail Rates

Had competition developed very rapidly, the distorted local rate structure evolved under state regulation would have come under substantial stress. But local competition has been slow to develop, as we have seen, and as a result there has been little or no movement in local retail rates since the passage of the Act (see Table 1.4).[12] In virtually every state, local rates have not changed perceptibly despite the fact that they do not reflect the incremental cost of service and therefore could not survive in a competitive marketplace. Thus one of the potential sources of large economic benefits – rate rebalancing – has been squandered for the first four years of implementing the 1996 Act.

Table 1.4 Average monthly business rates in urban areas, 1993–99 ($ per month)

	1993	1994	1995	1996	1997	1998	1999
Residential	19.95	19.81	20.01	19.95	19.88	19.76	19.87
Single-line business	42.57	41.64	41.80	41.81	41.67	41.28	41.00

Note: As at 15 October

Source: FCC (2000b)

5.4 Local Competition and the Cost of Entry

One of the major reasons for the slow pace of local entry and the limited use of leased loops is that the construction of local telecommunications networks requires substantial time and very large capital resources. Much of the investment is sunk cost, which causes enough uncertainty about the future to have a significant effect on investment incentives.[13] Furthermore, the FCC has determined that the wholesale prices of network elements should be set below the true economic cost, surely reducing the incentive for entrants to invest in their own networks, directly contrary to the goals of the 1996 Telecommunications Act. As we have shown, only one-third of CLEC lines is provided by their own network facilities.

Despite the favourable conditions created for CLECs by the availability of access to incumbents' facilities, the entrants have not been particularly successful. Not only have they failed to build networks that reach many of the dispersed small business and residential customers, but they have been encountering severe difficulties. Figure 1.2 shows the market capitalisation of all listed incumbent local exchange companies (ILECs), long-distance carriers (LDCs) and new entrants (CLECs). Note that the market capitalisation of the listed new entrants in the figure was $95 billion in mid-2000, three times that of the US airline industry. Since that time, however, CLEC share prices have collapsed and many of the new carriers have encountered severe difficulties. The result is that now the CLECs have less than $20 billion in market capitalisation. The decline has been almost as severe for the long-distance companies that have encountered increasing competition and downward pressure on their rates.

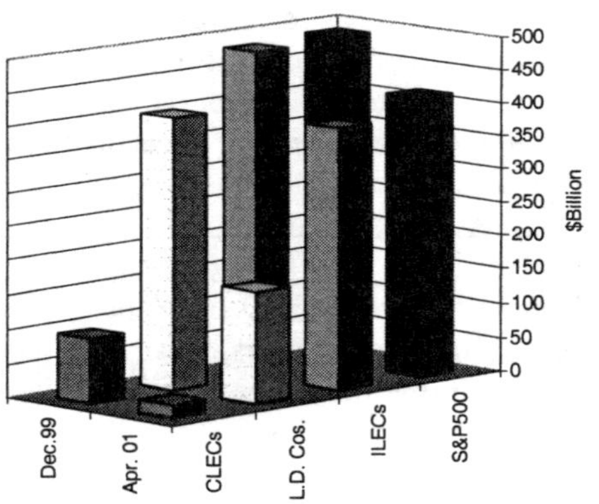

Source: Compiled by authors

Figure 1.2 Market capitalisation of three groups of US carriers, December 1999 vs April 2001

Many of the CLECs have not been able to obtain the financing required to continue expansion of their networks and to attract additional customers.

A number of CLECs have declared bankruptcy and their asset values have declined to very low levels. The regulatory approach, which has emphasised network sharing at below cost rates and resale, has contributed to the financial distress of the CLECs. The assets values of the CLECs are depressed in no small part because of the below-cost rates afforded them by regulation.

Facilities-based entry can easily require $1,000 or more per home passed for fibre-optics systems, including start-up losses. Since technology is changing very rapidly and a large share of this investment may be truly sunk, build-out plans must be cautiously conceived. Had the regulatory climate not sown confusion, discord, and innumerable court challenges, capital formation by new entrants might have been somewhat greater, but at this juncture it is not clear that many of the entrants have discovered a successful business strategy given the regulatory framework established under the 1996 Act.

5.5 Long-distance Competition

Until 1984, long-distance competition in the US was frustrated by continuing disputes between AT&T and the new entrants. The AT&T divestiture ended these disputes, and the AT&T decree's equal-access provisions placed entrants and incumbents on an equal footing with regard to access (interconnection) conditions and charges. In essence, the AT&T decree provided for long-distance entrants what the reciprocal compensation requirements of the 1996 Act provided for local entrants after 1996. The result of the 1984 equal access provisions on long-distance competition in the US appears to have been dramatic. As Table 1.5 shows, AT&T began to lose market share rapidly after 1984. Even though competition had begun in the mid-1970s, AT&T still had more than 90 per cent of long-distance carrier revenues a decade later. In the next 10 years, however, its share fell from 90.1 per cent to 55.2 per cent. In the first three years after the divestiture it lost approximately 12.8 per cent of its market share (11.5 per cent divided by 0.901), a far greater share loss than has occurred in the local markets in the first four years after the passage of the 1996 Act.

US long-distance rates continue to fall in the face of increasing competition and declining access charges. The FCC estimates that the average revenue per domestic interstate minute of long-distance service fell from $0.15 in 1992 to $0.11 in 1999 (Lande 2000, Table 10). About $0.03 of this decline is attributable to declining access charges collected by the local carriers and universal service charges imposed by the FCC. The remainder is a reflection of competition and increased productivity. Most of this decline appears to be concentrated in business rates, not residential rates. On average, residences paid about $0.175 in 1996 and 1997, and 14 cents in 1999 (TNS

Telecoms). Given that business calling accounts for much more than half of all long-distance calls, business rates averaged somewhat less than 8 cents per minute in 1999, still above the long-run incremental cost of service – access charges plus $0.015 per minute – or $0.045 per conversation minute (Lande 2000; Crandall and Waverman 2000). Residential rates remain far above the long-run incremental cost of physically delivering the service, creating the need for further market entry. Unfortunately, entry from the most likely sources – the Bell companies – has been blocked by continued disputes over the Bell companies' compliance with the detailed regulatory 'checklist' in the 1996 Act.

Table 1.5 Market shares of revenues for long-distance carriers, 1984–99 (%)

	AT&T	MCI	Sprint	WorldCom	All Others
1984	90.1	4.5	2.7	–	2.6
1985	86.3	5.5	2.6	–	5.6
1986	81.9	7.6	4.3	–	6.3
1987	78.6	8.8	5.8	–	6.8
1988	74.6	10.3	7.2	–	8.0
1989	67.5	12.1	8.4	0.2	11.8
1990	65.0	14.2	9.7	0.3	10.8
1991	63.2	15.2	9.9	0.5	11.3
1992	60.8	16.7	9.7	1.4	11.5
1993	58.1	17.8	10.0	1.9	12.3
1994	55.2	17.4	10.1	3.3	14.0
1995	51.8	19.7	9.8	4.9	13.8
1996	47.9	20.0	9.7	5.5	17.0
1997	44.0	19.0	9.5	6.6	20.9
1998	43.5	23.5	8.5	–	24.9
1999	40.7	23.7	9.8	–	25.7

Notes: *MCI and WorldCom merged in 1998*
Data in MCI column for 1998 and 1999 are for MCI/WorldCom

Source: FCC (2000a)

The forgone economic gains from blocking Bell-company entry into long-distance services are considerable. For example, assume that such entry had lowered interstate residential long-distance rates by 4 cents per minute in 1999 – still leaving them substantially above business rates. Assuming that residences account for 40 per cent of all domestic interstate minutes, this reduction would have generated consumer gains of $6.1 billion, including a transfer from producers of $5.4 billion and a reduction in dead-weight loss of

$0.7 billion. Given that intrastate long distance is about one-third of long-distance services, the consequent liberalisation of intrastate long distance could have generated another $3 billion in consumer gains.

6. THE WIRELESS SECTOR: COMPETITION WITHOUT REGULATION

Competition in US commercial wireless services had been slow to develop until very recently. For more than a decade the US had but two wireless (cellular) providers in each market because of an FCC decision to allocate only two 25 MHz licences to each market. In addition, unlike Europe or Canada, the US licensed wireless services on a geographically fragmented basis. In 1993, however, the Congress forced a major change in US wireless policy by instructing the FCC to auction spectrum, in part to solve the US government's budget deficit and essentially to deregulate cellular pricing. These auctions began in December 1994 and have resulted in the assignment of 120 MHz of additional wireless spectrum for 'personal communications services' (PCS) throughout the country. In addition, an entrepreneurial company, Nextel, has succeeded in transferring another 10–15 MHz of spectrum from a dispatch mode service to a commercial wireless service that competes with cellular and PCS service providers.

6.1 Wireless Rates

The first of the new PCS services began in late 1995, based upon the first two 30 MHz blocks of spectrum that were auctioned between December 1994 and March 1995. Since that time, a large number of companies have begun building facilities. All but 28 of the top 100 metropolitan markets in the US now have at least five wireless competitors – two cellular providers, two to four PCS services, and Nextel. The effect of the resulting competition on wireless rates in the US has been stunning. Throughout the 1984–95 period, real, inflation-adjusted cellular rates had fallen at a rate of 3 to 4 per cent per year (Hausman 1999c). Between 1995 and 1999, however, real cellular rates fell at a rate of 17 per cent per year as PCS service providers offered service at prices per minute that were more than 50 per cent lower than existing cellular rates. There is evidence that with open entry just one new player is sufficient to drive rates sharply lower. As the wireless industry evolves from its current structure into one with six or seven competitive, national players, there are likely to be further major beneficial effects on both rates and service quality.

The capital requirements for building these new PCS systems are considerable. The PCS providers have been spending between $1.5 billion and $2.2 billion per year on their infrastructure. The established cellular companies have been spending even more, upwards of $3 billion per year, to convert their infrastructure from analogue to digital. And both have been spending heavily to subsidise a portion of the annual $6 billion investment in consumer handsets (Leibowitz 1999). All told, wireless companies have been spending more than $12 billion per year over the past three years. Were this investment directed to wireline services, it could have resulted in providing facilities for serving perhaps 30 million subscribers, far more than the CLECs and CAPs currently serve. Competition in a very capital-intensive sector – the wireless sector – followed immediately upon the liberalisation of entry conditions. Subscriber rates fell dramatically, and they continue to fall as the new PCS companies fill out their networks.

The only new regulation imposed on the wireless industry by the 1996 Telecommunications Act was the imposition of 'reciprocal compensation' on wireless–wireline interconnection. Prior to 1996 these rates for exchanging traffic were often in excess of 2 cents per minute; today they are in the 0.5 to 0.7 cents per minute range because wireless companies are afforded the same interconnection rates as the new CLECs. Otherwise, competition has increased dramatically without unbundling, resale or other new forms of regulation. Indeed, wireless services are essentially fully exempt from retail price regulation. The result has been a major decline in real wireless prices.

Wireless rates have continued to decline, perhaps at an accelerating rate. In May 1998 AT&T introduced a flat-price nation-wide calling plan that allowed a subscriber to make calls to and from anywhere in the continental US for a single rate of $0.11 per minute to $0.15 per minute, depending on the volume of calling. Roaming charges were simply eliminated under this plan. The other carriers followed with $0.10 per minute plans, and Bell Atlantic even has one plan that offers nation-wide calling at only $0.0833 per minute. These sharp declines in the price of wireless calling are now sparking interest in developing wireless as an alternative to wireline service. Indeed, AT&T is promoting its service in Texas as the alternative to a subscriber's second line.

6.2 Subscriber Penetration

The US has lagged some European countries in cellular penetration, in large part because of its fragmented wireless industry and the need for subscribers to roam onto other systems at high prices. The development of a national service by AT&T, Sprint and Nextel, as the others – Verizon, Voicestream, and Cingular – expand their footprints, has led to national pricing without roaming charges. While subscriber penetration is still only about 40 per cent of the population, it is now growing rapidly.

As long as the US subsidises residential wireline connections, US penetration may remain below that realised by Hong Kong, Italy and the Scandinavian countries, but we have not yet witnessed the full effect of the sharp decline in US wireless rates. Nor has the US been able to implement 'calling-party-pays' tariffs, thereby freeing wireless subscribers to leave their lithium-ion powered handsets on continuously to receive calls.

6.3 Wireline-wireless Substitution

There is substantial casual empirical evidence that wireless service is a substitute for traditional wireline telephone service in several countries. Subscriber line growth has stopped in several Scandinavian countries and is likely to turn negative as wireless service becomes virtually universal. Some forecast that many European countries will have more than 70 per cent penetration of wireless service, meaning that virtually every adult will have a portable wireless handset (Mutschler and Wuh 1999). Wireless services may already be a weak substitute for wireline telephone service in the US, but the new low-priced bulk-pricing plans may be too recent to have induced a major shift from wireline to wireless service yet, particularly given the recent decline in consumer long-distance rates.

First, it is clear that heavy users of wireless services are also intensive users of wireline services, but, more importantly, their ratio of wireless to wireline expenditures increased between 1997 and 1998. Households generate large telephone bills through heavy use of out-bound long-distance services. The evidence in Table 1.6 thus shows that as their long-distance calling increases, their use of wireless services also increases. Moreover, the ratio of their wireless expenses to their wireline expenses appears to be rising over time despite declining long-distance rates because wireless rates have been falling even more rapidly. These patterns are suggestive of a growing substitution of wireless for wireline services. As real wireless prices continue to fall and the real cost of residential connections rises due to the rebalancing forced by local competition, we should see a substantial increase in the usage of wireless service from its current 6 per cent of total minutes of use in the US.

There is another form of wireless telecommunications that could compete with wireline telephony, particularly in areas of low population density. 'Fixed' wireless service, which connects fixed subscribers to the network through signals transmitted through the radio spectrum from terrestrial towers, is already in use in many developing countries in areas of low density. It could be deployed to a much greater degree in advanced countries such as the US but for 'universal service' pricing policies that keep rural rates below the long-run incremental cost of the wireline service. Were current telephone rates set efficiently, such wireless services might develop rather rapidly and provide substantial potential and actual competition for even the lightly-populated areas of the country. Unfortunately, the 1996 Telecommunications Act allows states to exempt rural areas from competitive entry. Therefore, even if current rates were rebalanced to economically efficient levels, or if explicit, transferable subsidies were imposed to cover access deficits in rural areas, many states might still refuse to allow rural areas to enjoy the benefits of competition.

Table 1.6 Average household bills for most intensive users of wireless, 1997 and 1998

Wireless spending percentile	1997 wireline	1997 wireless	1997 ratio wireless/ wireline	1998 wireline	1998 wireline	1998 ratio wireless/ wireline
70 – 75	84.73	47.36	0.56	80.85	46.72	0.58
75 – 80	87.65	53.78	0.61	88.27	53.13	0.60
80 – 85	92.03	62.81	0.68	90.58	62.18	0.69
85 – 90	101.60	77.11	0.76	95.54	75.17	0.79
90 – 95	122.31	100.86	0.82	108.18	98.08	0.91
95 – 100	133.37	223.63	1.68	123.03	194.22	1.58

Source: TNS Telecoms

6.4 Payphone Use

Decreased wireless prices and increased wireless penetration have already had a significant effect on payphone usage. The number of payphones has decreased by about 17 per cent over the past three years. With integration of networks in the future, including voice and data networks to allow easy conversion from e-mail to voice mail, and the ability to view e-mail on hand-held mobile devices, a mobile telephone will likely become the chief instrument for voice messages in the future. Especially in a business setting, the ability to avoid the frustration of endless voice mail messages is likely to lead to significant substitution of mobile voice for current wireline voice usage. One detriment to even high mobile usage is the relatively high tax rates levied on mobile usage in the US by a combination of federal, state and local tax rates (Hausman 2000).

7. REGULATION AND INVESTMENT

As we have seen, the new 1996 Telecommunications Act provides extensive new regulation of the nation's existing local exchange carriers, but leaves the wireless sector to develop on an unregulated basis. In addition, the 1996 Act provides a substantial modicum of rate deregulation for the cable television industry and allows cable companies to provide traditional and new advanced telecommunications services free of the wholesale and retail regulation that applies to their rivals in the local exchange sector. All three groups – the local exchange companies, cable television, and wireless service providers – are important potential sources of new technologies and facilities for delivering broadband internet connections. Indeed, AT&T has invested more than $100 billion in cable systems as part of a strategy to deliver the full range of video, data and voice services to residential customers. The local exchange companies have begun to deploy their own broadband service over their copper wire distribution lines, called digital subscriber line (DSL) service. And wireless companies continue to invest in new digital technologies capable of offering ever greater bandwidth and features.

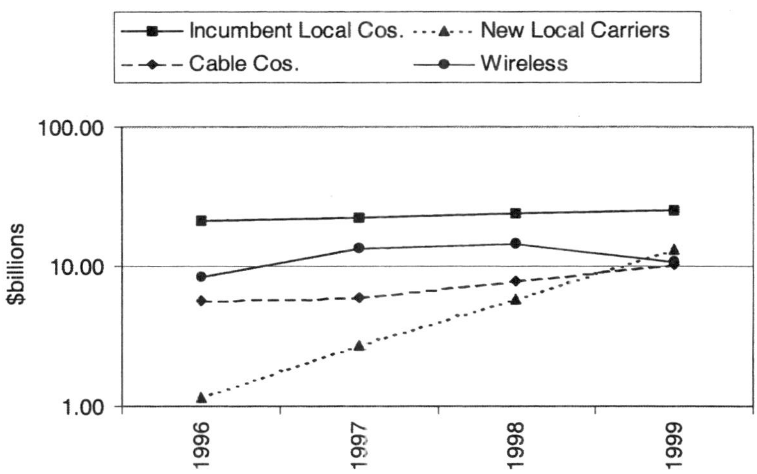

Source: Cellular Tecommunications Industry Association (drawn from annual reports of member companies)

Figure 1.3 Capital expenditures in the US, 1996–99

It is still too early to estimate the effects of the regulatory climate on investment strategies in any detail. However, a brief look at recent investment trends suggests that regulation may be a substantial drag on investment in the traditional telephone industry. Despite the very large potential market for new, advanced broadband services for internet connections, the heavily regulated incumbent local exchange companies have increased their capital spending only modestly in recent years. At the same time, capital spending by the cable/broadcast sector and by wireless companies, which essentially are free of price regulation, has soared, although the latter has levelled off since 1998 (see Figure 1.3). The new entrants into local telephony invested heavily in 1997–99, but their subsequent financial problems will undoubtedly lead to sharp reductions in capital spending. The slow growth in total capital spending by the local incumbents in a sector that has recently been opened to competition should send off warning alarms that something is amiss. This difference in the investment behaviour of the regulated and the unregulated sectors of the communications industry provides further evidence that FCC regulation of new services and FCC-state regulation of wholesale and retail rates is probably retarding capital formation.

8. LESSONS FOR OTHER COUNTRIES

Competition in US telecommunications markets is only beginning in the local telephone access/exchange market. The complex regulatory structure developed by the US to manage local competition has been extremely controversial and has created years of regulatory strife and litigation. 'Regulated competition' may well be an approach that harms consumers while regulators attempt to help new competitors. By relying upon a return to a flawed system of cost-based regulation to provide entrants with extremely favourable access to incumbents' facilities, regulators are attempting to micro-manage the liberalisation process in a manner that is quite unprecedented in the recent history of US deregulation. Moreover, there is very little evidence that this management of competition is working very well. Rather, this regulatory approach becomes mired down in a hopeless exercise of setting wholesale rates and limiting incumbents' ability to exploit the value of their own networks. This approach stunts the existing carriers' incentives to invest in new facilities and delays investment by the entrants as they wait for regulators to provide them with access to the full complement of incumbents' facilities at below-cost prices. After five years the competitors have captured substantially less than 10 per cent of subscriber lines.

By contrast, competition in the partially deregulated long-distance market has increased, and competition in the totally deregulated wireless market is flourishing. Long-distance rates, particularly business rates, are falling steadily. Residential rates are still substantially above long-run incremental cost, pointing out the need for further entry. Wireless rates are falling even more dramatically. New wireless pricing plans are emerging and wireless carriers are experimenting with new services in a totally deregulated market. It appears likely that wireless will soon be a sufficient source of competition for conventional voice/data wireline services to permit a substantial degree of deregulation of all telecommunications service providers, but it is unlikely that such deregulation can or will occur, given the detailed mandates of the 1996 Act. Nor is it likely that a reconsideration of the 1996 Act by Congress would lead to better legislation and an improvement in the prospects for deregulation. Other countries should surely consider wireless/wireline competition as a substitute for regulation.

A recurring theme in this chapter is that detailed cost-based regulation of wholesale rates is not a satisfactory approach for stimulating competition in this network industry, particularly one in which technology is changing rapidly. The current cost-based regulatory regime in the US has not encouraged entry to any meaningful extent, but it is discouraging investment in new facilities. Given the large amounts of capital required to develop new 'broadband' services for high-speed internet connections, this effect is particularly unfortunate. Other countries, particularly those still requiring substantial investment in basic telecommunications infrastructure, would be well advised not to emulate the intensely regulatory approach adopted by the US.

We would prefer an attempt by regulators to undo the regulatory-created barriers to entry built into the retail rate structure. As long as large numbers of subscribers – particularly residential subscribers in all but the most dense areas of the country – are provided local service at rates that are below long-run incremental cost, entrants will have little incentive to offer service to these subscribers. It would be better for regulators to attempt to eliminate the pricing distortions that they have already created than to erect an entire new set of distorted wholesale rates that will be difficult to eliminate in the future. An ounce of regulatory forbearance is surely much to be preferred over a pound of regulatory cure.

NOTES

1. The Commission had instituted price-cap regulation – a regulatory regime that limits the rate of annual price increases for various groups or 'baskets' of services – in order to divorce rate increases from cost increases. This change was instituted to simplify the regulatory process and to provide greater incentives for carrier efficiency since reductions

in a carrier's costs would not require a reduction in regulated rates. Favourable effects on technological progress and innovations were also an expected result of the shift to price caps.

2. For an analysis of this competitive entry see Hausman (1996).
3. Note that these cost models estimate the average incremental cost of providing a local service in a given area, not the long-run incremental cost of adding a line in any given area. Note also that in Figure 1.1 the 'BCPM' (benchmark cost proxy model) estimate is taken from a local carrier model proffered to the FCC, while the 'HCPM' (hybrid cost proxy model) estimate is derived from the FCC's own forward-looking cost model.
4. Estimates are based on Crandall and Waverman (2000).
5. The FCC approach to setting wholesale rates does not take account of the sunk and irreversible costs of investment in telecommunications networks. For a further discussion see Hausman (1997; 1999a; 1999b).
6. This is somewhat overstated. The ILECs faced competition from urban fibre rings and wireless carriers before 1996, but they controlled virtually the entire market for local access and exchange services for residential and small business subscribers.
7. For a discussion of the new subsidy programmes and 'universal service' policies see Crandall and Waverman (2000). For a discussion of the increased taxation from these policies and the extremely large loss of economic efficiency and reduction in consumer welfare caused by the FCC's implementation of these policies see Hausman (1998) and Hausman and Shelanski (1999).
8. For a consumer welfare approach to unbundled elements, rather than the FCC's competitor welfare approach, see Hausman and Sidak (1999).
9. This discussion is based on Hausman (1997; 1999a; 1999b).
10. One of the founders of the TSLRIC (which mutated into the TELRIC) approach has now recognised that significant costs are omitted by this approach (see Baumol 1999).
11. The shares shown in Table 1.3 reflect the RBOCs' shares of total lines for the entire state, not just the dense, urban markets. Most states have a single UNE rate for loops across the state.
12. In earlier exercises of 'deregulation' there was often substantial volatility and a sharp decline in prices charged by the erstwhile regulated carriers (Winston 1998). Because competition has developed slowly in local telecommunications and retail rates remain regulated, this volatility has not yet emerged in US telecommunications.
13. See Dixit and Pindyck (1994) for a further discussion of the effect of uncertainty and sunk costs on investment incentives.

ABBREVIATIONS

AIC	average incremental cost
AT&T	American Telephone and Telegraph Company
BCMP	benchmark cost proxy model
CAPs	competitive access providers
CLECs	competitive local exchange carriers
DSL	digital subscriber line
FCC	Federal Communications Commission
GDP	gross domestic product
HCMP	hybrid cost proxy model
ILECs	incumbent local exchange companies
LATAs	local access and transport areas

LDCs	long-distance carriers
PCS	personal communications services
RBOCs	regional Bell operating companies
TELRIC	total element long-run incremental cost
UNEs	unbundled network elements

REFERENCES

Baumol, W. (1999), 'Option value analysis and telephone access charges', in J. Alleman and E. Noam (eds), *Real Options: The New Investment Theory and its Implications for Telecommunications Economics*, Dordrecht: Kluwer, pp. 215–18.

Crandall, R.W., and L. Waverman (2000), *Who Pays for Universal Service? When Subsidies Become Transparent*, Washington DC: Brookings Institution.

Dixit, A. and R. Pindyck (1994), *Investment Under Uncertainty*, Princeton, NJ: Princeton University Press.

FCC (2000a), *Statistics of the Long Distance Telecommunications Industry*, Report, Washington, DC.

FCC (2000b), *Trends in Telephone Service*, Report, Washington, DC.

FCC (2001), Local Telephone Competition: Status as of 31 December 2000, May, Washington DC.

Hausman, J. (1996), 'Proliferation of networks in telecommunications', in D. Alexander and W. Sichel (eds), *Networks, Infrastructure, and the New Task for Regulation*, Ann Arbor: University of Michigan Press, pp. 19–36.

Hausman, J. (1997), 'Valuation and the effect of regulation on new services in telecommunications', *Brookings Papers on Economic Activity: Microeconomics*, Washington DC, 1–18.

Hausman, J. (1998), 'Taxation by telecommunications regulation', *Tax Policy and the Economy*, **12**, 29–48.

Hausman, J. (1999a), 'Regulation by TLRIC: economic effects on investment and innovation', *Multimedia Und Recht*, **3**, 22–6.

Hausman, J. (1999b), 'The effect of sunk cost in telecommunication regulation', in J. Alleman and E. Noam (eds), *Real Options: The New Investment Theory and its Implications for Telecommunications Economics*, Dordrecht: Kluwer, pp. 191–204.

Hausman, J. (1999c), 'Cellular telephone, new products and the CPI', *Journal of Business and Economics Statistics*, **17** (2), 188–94.

Hausman, J. (2000), 'Efficiency effects on the US economy from wireless taxation', *National Tax Journal*, **53** (2), 733–42.

Hausman, J. and H. Shelanski (1999), 'Economic welfare and telecommunications welfare: the e-rate policy for universal service subsidies', *Yale Journal on Regulation*, **16** (19), 20–51.

Hausman, J. and G. Sidak (1999), 'A consumer-welfare approach to mandatory unbundling of telecommunications networks', *Yale Law Journal*, **109** (3), 417–506.

Janney Montgomery Scott Inc. (1997), *The Status of Agreements Between the Major LECs and CLECs: Update*, 10 November.

Lande, J. (2000), *Telecommunications Industry Revenue 1999*, Washington DC: Industry Analysis Division, Common Carrier Bureau, FCC.
Leibowitz, D.H. (1999), *The Wireless Communications Industry*, New York: Donaldson, Lufkin and Jenrette Securities Corp.
Mutschler, L.J. and P. Wuh (1999), *The Next Generation III: Wireless in the US*, Merrill Lynch: New York, 10 March.
TNS Telecoms (various), *Bill Harvesting Data*, Jenkintown, PA.
Winston, C. (1998), 'US industry adjustment to deregulation', *Journal of Economic Perspectives*, **12**, Summer, 89–110.

2. United Kingdom: Economic and regulatory issues of telecommunications reform

Cosmo Graham and Moazzem Hossain

1. INTRODUCTION

Developing a prosperous and vibrant telecommunications industry is critical for any country's economy today. This is not solely due to the productivity of the industry itself, important as that may be, but because it is a platform for a multitude of modern enterprises. Efficient telecommunications are vital to the development of 'e-businesses', as well as many others, and are also important if a country wishes to develop itself as an 'information society'. Particularly in an era of globalisation, where so much depends on a country's ability to communicate with other parts of the world, telecommunications are vital.

In recent years, economic reforms through liberalisation and privatisation, initiated in the USA and the UK in the 1980s and later spread to many other countries through, among other agents, the World Bank and the IMF (International Monetary Fund), have become major vehicles for achieving economic growth in both developed and developing nations. This looks like a particularly attractive approach in relation to telecommunications, as the public sector cannot provide the sums of money, on its own, necessary to drive such a dynamic industry forward. Therefore, a country will have to create the necessary conditions that will make it attractive to investors. However, these reforms are not sufficient on their own to ensure success. Levy and Spiller's (1994) work for the World Bank suggests that a critical element for the success of privatisation and liberalisation is a stable and predictable regulatory system, which also implies an open regulatory system. Although there are a number of ways of accomplishing these ends, in this chapter we examine the UK regulatory experience in order to see what lessons it offers for the Asia–Pacific developing nations. We have chosen the UK because it was the first country to privatise its telecommunications industry and thus there is a long historical record to look back upon. A word of caution is necessary, however, because the UK's experience is notably different from that of other developed nations, partly due to different constitutional arrangements, partly because it has moved very fast towards full privatisation of utilities and because, in relation to telecommunications, it has attempted to create competing infrastructures, rather than competition in services. The present chapter analyses the economic, legal and regulatory issues of the telecommunications reform in the UK. It includes six sections: the next section discusses the privatisation process in the UK. Section 3 outlines the legal structure and certain changes which have been undertaken; section 4 addresses the developments in market structure and the economic issues of telecommunications reform; section 5 outlines some recent regulatory issues and the final section attempts to draw some lessons from the UK experience for the Asia–Pacific developing nations.

2. PRIVATISATION OF TELECOMMUNICATIONS IN THE UK

2.1 Privatisation

The decision to privatise the then British Telecommunications (hereafter BT) in 1984 was, politically, quite extraordinary. Although there had been previous privatisations, nothing of this scale had been attempted before (generally see Pitt 1990 and Newman 1986). Although one cannot be sure until the historical records are opened up, the decision seems to have been due to a number of factors that combined to make this seem the best option. The Conservative government had a preference for market solutions and a desire to 'roll back the state' or, more prosaically, a wish not be involved in the running of what they saw as commercial enterprises. A critical part of this platform was a desire to reduce the burden of direct taxation on voters, but this depended on reducing government borrowing. Due to the arcane rules of public sector financial accounting in the United Kingdom, the privatisation of nationalised industries directly reduced the figure for government borrowing, thus making it easier to cut taxes.

This combined with a general policy difficulty, of how exactly the state should control nationalised enterprises, which had never been satisfactorily resolved. There was also a more specific issue, as far as BT was concerned, which was that it needed to raise substantial funds for investment, which it was unable to do within the public sector. Just to add icing to the cake, the possibility of creating millions of small shareholders was also perceived as electorally advantageous to the Conservative party, which had seen the political benefits of privatising council houses.

The sale of 51 per cent of the shares of BT in November 1984 raised a total of £3.9 billion, six times larger that any previous issue on the United Kingdom Stock Exchange (Bishop and Kay 1988, p. 4). Given that it was such a large flotation, that the sell-off was critical to the government's financial needs and political aspirations and that earlier privatisations had had mixed results in terms of flotations, it is not surprising that the government adopted a cautious approach to the matter of increasing competition (Moon et al. 1986). Although there had been some liberalisation in 1981, primarily of telecommunications apparatus, and the government's economic adviser had urged unrestricted resale of voice telephony (Beesley 1981), a more cautious approach was taken. BT was privatised intact and initially only one competitor, Mercury Communications, was allowed. (This was known as the duopoly policy.) Simple re-sale of telephone services was not allowed. There has been criticism, both specific to BT's sale and in relation to other privatisations, that such a cautious approach reflected pressures from the incumbent management who were naturally keen to maintain as much economic power as possible. It has been argued to the contrary that the government had gone for a limited measure of competition and that no one had studied the real problems of interconnection that would have followed from a greater degree of break-up and more competition, problems that have proved, as we shall see, very difficult to solve (Foster 1992; Lawson 1992).

There are two interrelated points here. The first is that, in order to privatise a telecommunications industry successfully, it may be necessary to compromise on the degree of competition that is introduced, at least initially. Second, whatever bargain is struck at this level of the process will have important effects on the future regulatory system making it more or less difficult for the regulator to carry out their tasks.

3. TELECOMMUNICATIONS REGULATION IN THE UK

3.1 Legal and Institutional Structure

In this section of the chapter we discuss the regulatory system created in 1984 by the Telecommunications Act (TA), which still operates today, although it will soon be replaced by a new Office of Communications. The major regulatory innovation was the creation of an administrative agency, operating at some distance from government, the Office of Telecommunications (Oftel) headed by a Director General, appointed by the Minister, who had formal legal responsibility for the operation of Oftel's powers. Provision of telecommunications services was made subject to a licensing regime, that is, the provision of telecommunications was a criminal offence unless carried out under a licence granted, not by Oftel, but by the Department of Trade and Industry (DTI). So the institutional arrangements were that commercial operations in the market were carried out by the licensees, (BT and Mercury initially), regulation was carried out by the Director General of Telecommunications (DGT) and the government took a back seat, with the exception of its licensing power. Under the legislation the government has no power to issue directions to the DGT, except in very limited circumstances (TA 1984, s. 94). There is a power to suggest formally what priority the DGT should take in relation to his or her duties, but this has never been exercised, although there have been informal contacts and the DGT has felt constrained by government's views and policies (TA, 1984, s. 47).

3.2 Duties of the Regulator and the Secretary of State

The legislation imposes general duties on the Director General, and the Secretary of State, which they are to take into account when exercising their functions, with the exception of dispute settlement, when they become objective interpreters of the rules. The role of the Director General and Secretary of State is not to represent any particular constituency but to act in a way 'best calculated', in their opinion, to achieve various objectives. This does not mean that they are under an obligation to balance conflicting interests, although they may come to the conclusion that the best way to achieve the objectives is to strike a balance between different parties, particularly if the statutory duties do not all point in one direction and the interests of the various parties conflict. They may equally come to the view that, in the light of the statutory duties, one interest should take priority over another.

The duties of the regulator and Secretary of State are set out in two sub-sections of the Telecommunications Act (TA 1984, s. 94). The duties in the first sub-section are the *primary* duties because the duties listed in the second sub-section are expressed to be *subject to* the duties in the first sub-section and are therefore *secondary* duties. The two main primary duties are: to ensure that all reasonable demands for the product are satisfied and that the companies concerned are able to finance the proper carrying out of their activities. A number of social concerns (emergency services, public call box services, directory information services, maritime services and services in rural areas) are expressed as primary duties for the regulator but they are made subject to the financing duty. What does this entail?

The first point is that the financing duty does not entail giving priority to the shareholders' interests, as opposed to the consumers' (Prosser 1997, pp. 20–1). If the companies are unable to finance their activities, there will be no services for the consumers. In fact, this provision has to be interpreted alongside the duty to satisfy reasonable demands. In the telecommunications legislation the Director General, when considering this duty, must do so without prejudice to the general duty to ensure all reasonable demands are met. The legislation envisages that the demand for services may be reasonable but that such demand may cause problems for the financing of the service providers, in other words, that there might be a tension between the two limbs.

The duty to satisfy all reasonable demands may be interpreted as a duty of universal service. This interpretation is supported by the existence of a duty to supply (BT licence condition 1) in certain circumstances being placed upon the regulated companies that has been interpreted by the courts as leaving no discretion on the part of the supplier when the stipulated conditions are met.[1] If this is the case then, at the heart of the legislation is an acknowledgement that telecommunications services are, in some sense, different from ordinary commodities, although similar to electricity, gas and water, because there is no duty on any ordinary business to supply its goods or services to anyone who reasonably asks for them.

Outside of these primary duties there are a number of secondary duties amongst which is the protection of consumers, both domestic and business, and an argument has been put forward, not least by the regulators (Vass 1995 and Oftel 1997a, p. 19), that consumer protection is their primary rationale. A straightforward reading of the legislation does not support this but the argument runs as follows: the primary duties are to ensure that reasonable demands are met and that companies can finance their activities. It is in the consumers' interests that companies can finance their activities, otherwise no services can be provided. By ensuring that reasonable demands are met, the regulator will clearly be protecting the consumer interest. 'Reasonable' in this instance can be equated with the sort of demands that would be made if there was a competitive market. Since regulation is always a second-best solution, competition should be encouraged, which is in the consumers' interest because they then have real choices, they do not depend on a regulator mimicking market processes.

This is too simplistic an argument, given the terms of the legislation, as consumer interests are not given priority over any other duties in the legislation. For example, the Telecommunications Act contains a number of what might be called industrial policy duties. For example, enabling providers of telecommunications services to compete effectively outside the United Kingdom (TA 1984, s. 3 (2) (g)) pulls in a different direction from the duty to promote competition and the interests of consumers. It has in the past been assumed that one way of allowing a company to compete abroad is to provide them with a secure home market and cash-flow, although there is an alternative view that companies are more likely to become effective with a competitive home market. Most seriously, insofar as there is a clash between the companies' ability to finance their activities and the consumer interest, the former must prevail. This is not simply a choice between bankruptcy or not; companies will legitimately need to raise finance for a variety of projects in the short, medium and long-term. Given the demands of the British capital markets, companies might well find difficulty in financing legitimate projects if the consumer interest overrode, or arguably, was given an equal status.

As well as these industrial policy duties there is also a secondary duty to promote and maintain effective competition in relation to telecommunications in the United Kingdom. It is ironic, given the pace of change and the amount of competition in telecommunications, that this is only a secondary duty whereas in relation to the energy industries, which have been historically less competitive, this became a primary duty in the legislation up until the duties were re-cast in the Utilities Act 2000.

The point to take away from this discussion of the duties is simple. As the Better Regulation Task Force (2001, p. 6) put it, '[the statutory objectives] can be contradictory and it is difficult for companies to judge how regulators will prioritise between them.' As will be evident from the discussion above, this is inevitable given the nature of the policy concerns. Relatively greater weight should be paid therefore to the institutions, which manage these tensions and the procedures by which they take policy decisions.

3.3 Functions and Responsibilities of Regulators

3.3.1 Responsibilities in relation to licences – grant, modification and enforcement

The central instrument of the regulatory system is the licence granted to the utility companies.[2] The licence sets out, in a public document, the terms and conditions under which the utility company will operate and it is the responsibility of the regulator to police the terms of the licence and to suggest modifications. The original BT licence had 53 conditions and was over one hundred pages long.[3] The key conditions were: controlling prices, certain social obligations, such as universal service, conditions relating to the supply of services, use of the company's network by competitors and provisions preventing unfair discrimination. There were also conditions on the provision of information to the regulator and the need to produce Codes of Practice for relationships with consumers. A similar licence was granted to Mercury, although without the price control. Note that the legislation assumes that licences are individualised documents, that they are not standard documents applying to a class of persons. The UK regulator has no general rule-making powers. If the regulator wishes to make an industry-wide change then this must be done either through modifying all the relevant licences, one by one, or by convincing the government to pass legislation.

Once a licence is granted it is very difficult to change without the consent of the licensee. Thus a key question is who grants the original licences and under what terms. In telecommunications, the Secretary of State, not the regulator, granted the original licences. In other words, the government retained formal control over the development of competition.

Once a licence is issued it is part of the Director General's duty to keep under review the functioning of the relevant industry and, if necessary, propose changes to the licence (TA 1984, s. 47 (1)). If a Director General wishes to modify a licence there are two procedures that can be followed. First, if the company agrees, the Director can modify the licence subject to following certain procedural requirements relating to publicising the proposed modification and considering any representations or objections. The Secretary of State has the right to veto any proposed modification that might be agreed between the company and the Director General (TA 1984, s. 12). If agreement cannot be reached with the company, the only option for the Director General, if he or she does not want to abandon the idea of a licence modification, is to make a reference to the Competition Commission (CC) asking it to report on whether or not certain matters operate, or may be expected to operate, against the public interest and, if they do, could those effects adverse to the public interest be remedied or prevented by modifications to the licence (TA 1984, s. 13). The Secretary of State can only block the reference by the Director to the CC when matters of national security or international relations become relevant (TA 1984, s. 13 (5)). After a report the Director has a discretion to make such modifications to the licences as appear to him or her to be suitable for the purpose of remedying the adverse effects specified in the report, subject to having regard to the modifications specified in the report (TA 1984, s. 15).

The combined effect of these provisions is to impose substantial costs on the regulator and the licensee if he or she wishes to modify a licence over the objections of the company as they will have to present a case to the CC and this will take time and money and the CC may not agree with the regulator. The result is that both the regulator and the licensee will be encouraged to bargain, in the attempt to obtain an agreement, as well as having to ensure that the Secretary of State will accept any agreed licence modification. The process of reference to the CC has only been utilised three times (MMC 1989, 1995 and 1999) and one of these was by mutual consent.

In addition to modifying licences, the Director General is also responsible for their enforcement. If the Director is satisfied that a licensee is contravening or has contravened and is likely again to contravene any of the conditions of the licence he or she may make a final or provisional order in order to secure compliance with the licence conditions. Before making a final order or confirming a provisional order, the Director must follow certain procedural requirements which, in essence, require the company, and other parties, to be given notice of the charges and the opportunity for a hearing. If the company is still dissatisfied with the order it may apply to the court within 42 days of the order and seek to have the order quashed on the grounds that it was not within the powers of the Director General or that there has been a breach of the procedural requirements which has substantially prejudiced the interests of the company (TA 1984, ss. 16–18). Note that there is no power in the Telecommunications Act to fine a company for breach of its licence conditions.

3.3.2 Competition law powers and relationship with the competition authorities

In addition the DGT had concurrent powers with the Office of Fair Trading (OFT) under the competition legislation in relation to monopoly situations and anti-competitive practices, although not mergers, which remain in the hands, ultimately, of the Secretary of State. In 1984 these competition law powers suffered from the usual deficiencies of United Kingdom competition law, namely that the DGT would only have power to investigate and, if necessary, refer the matter to the CC. In lieu of a CC reference, the DGT could accept undertakings. There were no powers to fine businesses, nor were there any powers to issue 'cease and desist' orders. The only time a Director could take formal action was if there was a breach of a licence condition.

Although United Kingdom competition law powers were weak, BT was also subject, insofar as it engaged in inter-state trade, to European Community (EC) rules on competition law contained in Articles 81 (previously Article 85) and 82 (previously Article 86) which prohibited anti-competitive agreements and concerted practices and abuse of dominant position. These did make provision for fines of up to 10 per cent of the turnover of a company by the EC and the possibility of private litigation, although they have rarely been used.[4] In addition, since 1991, the EC has also had the jurisdiction to decide on proposals for mergers that have an appropriate Community dimension.

Because of the weakness of these competition law powers, and the lack of a power to fine BT for breach of its licence conditions, Oftel attempted to rectify this through inserting what it called a fair trading condition into BT's licence in 1995. The policy process which was followed is now largely of historical interest because the provisions have been overtaken by the Competition Act 1998 which gives the DGT powers to fine companies which are in breach of the prohibitions against anti-competitive agreements and abuse of a dominant position, which are modelled on Articles 81 and 82.

3.3.3 Quality of service and collection and provision of relevant information

One of the general functions of the DGT is to arrange for the publication of information, which he or she feels, is expedient to give to consumers and other users of the relevant services in their industry (TA 1984, s. 48). Perhaps surprisingly, the DGT was given no powers in the Telecommunications Act to make rules relating to the quality of service provided by telecommunications providers. The Director General's lack of a power to set standards for the quality of service soon proved to be a particular problem as, shortly after privatisation, BT stopped publishing its indicators of service performance. In order to obtain some information, the DGT had to organise his own quality of service surveys and, after putting considerable pressure on BT, he was able to gain the appropriate information from them to start publishing annual quality of service reports. The regulator for the gas industry had similar problems and, in regard to the later privatisation, the Electricity and Water Acts of 1989 provided the regulators in those areas with greater powers. Then in 1992 the Competition and Service (Utilities) Act made the legislation consistent across the four utility sectors, in line with an announcement made in the *Citizen's Charter* (Cm. 1599).

The current legal position is that the DGT has the power: to make regulations providing for standards of performance in individual cases which may include provisions for compensation if the standard is not met (TA 1984, s. 27A) set overall performance standards (TA 1984, 27B), publish information about the companies' performance in relation to individual and overall performance standards (TA 1984, ss. 27C, 27D), approve companies' complaints procedures (TA 1984, s. 27E) and to determine certain disputes over individual standards, discrimination in fixing charges, billing disputes and deposits. However, the provisions regarding billing disputes in telecommunications and the other industries have never been implemented (Bailey and Carney 1995) (see TA 1984, ss. 27A, 27F, 27G, 27I). Nor has the DGT utilised the legislative powers relating to quality of service, preferring to rely on voluntary standards originally negotiated with BT. The current position is that Oftel has encouraged the fixed lines and mobile phone industries to publish a set of six monthly comparative performance indicators.[5] The argument is that this is a better way, in a competitive market, of ensuring the appropriate standard of service, as opposed to Oftel specifying the requirements from on high.

For the sake of completeness, the DGT also has powers to investigate complaints between telecommunications providers and their customers, although, unlike other utility industries, there is no power to make a binding decision (TA 1984, s. 49). This is presumably because the relationship between telecommunications companies and their customers is now contractual, rather than statutory. Customers therefore have, ultimately, a remedy in the courts, although BT also operates an arbitration scheme for complaints. As far as consumer complaints are concerned, they have been a relatively low priority for Oftel, which has concentrated on informal negotiation to try and resolve them (Hall et al. 2000). However, the issue has recently become more important, as the numbers of complaints to Oftel have been rising, and Oftel has begun to search for ways to improve the companies' handling of complaints which has culminated in the recent establishment of a telecommunications Ombudsman (Oftel June 2000, p. 24 and July 2001a).

To summarise, in terms of domestic law, and with perfect hindsight, a number of problems and issues with the original legal structure can be identified. There was no provision for standard licence conditions or rule making. The problem for Oftel has been that, if it wishes to modify the provisions of a licence, the legislation provides that this can only be done with the consent of the individual licensee. Thus, when Oftel wished to incorporate a particular condition in all the other telecommunications licences it was forced to undergo an unwieldy, and artificial, consultation process in order to change the licences. By comparison, in both electricity and gas provision has now been made for altering standard conditions in licences by, in effect, a majority vote of the licensees. The DGT's enforcement powers were cumbersome and limited, as he could not impose monetary penalties and powers since domestic competition law were weak. In addition, there was no power to specify the quality of service requirements. On the institutional side, although possessing few formal powers with the exception of the critical power to issue licences, the Secretary of State's role was important.

4. ECONOMIC ISSUES OF REFORM

The economic issues of telecommunications reform in the UK have been dealt with in the literature extensively (see Armstrong et al. 1995; Meek 1998). In this section we examine three main issues: the question of market structure and the development of competition, the working of price control and the relative economic success, or otherwise, of BT.

4.1 Market Structure

The government was very cautious as regards changing the market structure after the privatisation of BT in 1985. Instead of allowing access by competitors to BT's network, allowing resale or breaking BT up into competing units, the so-called 'duopoly policy' was implemented. Only one national competitor to BT was licensed, Mercury Communications, which undertook to build its own network in order to compete with BT, as well as being given access to BT's network for its subscribers. The issue of interconnection was highly controversial and resulted in Mercury taking the DGT to court in a dispute that was ultimately resolved out of court. Nevertheless, the duopoly policy lasted from 1984 until 1991. A review was conducted jointly by the DTI and Oftel between the end of 1990 and early 1991 and was, apparently, subject to much lobbying by interested parties (DTI 1990, 1991; Thatcher 1994). The key results of the review were that the market was to be opened to other telecommunications operators but that BT was still to be barred from providing entertainment services for a period of ten years, although the DGT could review this after seven years. Although this looked like a defeat for BT, certain other matters were left to be elaborated by the Director General, notably the terms on which other telecommunications operators could interconnect with BT and what contributions they would have to make to finance BT's universal service obligations.

The decision to relax the duopoly policy opened the door for other competitors to BT's fixed link service. As a result, many network operators and service providers entered into the UK market post-1991. By 1997, Oftel's register of licences issued indicated 'there were over 180 fixed-link network operators in the UK. Over 100 of these were cable companies offering telephony services in franchised areas. The remainder of network operators included 30 national long-distance network operators and 50 international facilities-based operators' (Gist 1999, p. 123). In addition, post-1991 there were two other developments which became important for the development of the industry: the growth of cable TV companies which, almost incidentally, now offered telephone services and the growth of the mobile phone industry which started with the allocation of two licences in 1985 and a further three in 1991, reduced by consolidation to two. In what follows, we treat the mobile and fixed link markets separately, consistent with Oftel's latest analysis and for ease of presentation (Oftel July 2001b, paras 2.25–2.30).

Table 2.1 gives information on market share of the fixed link market between 1992 and 2000, using connections as measure. We can see from this that, although BT's market share on this measure has decreased somewhat, it is still by some distance the dominant supplier. If, however, we examine a different measure, that of call revenues and look at the statistics for the second quarter of 2000–01 (from Oftel's revenue data), BT has just under 74 per cent of the revenues from residential calls and just under 52 per cent of the revenue for all business calls. Even in the business sector, the next largest company had a market share of just over 15 per cent of business revenues. Oftel concluded on the basis of this information that, although BT remained the dominant company in all residential markets, in the business markets it was dominant only in local calls and calls to mobiles (Oftel July 2001b, paras 3.64–3.65). They concluded that, although BT was still dominant, there was evidence that competition was increasing.

Table 2.1 Residential and business connections and shares of main operators, 1992–2000

Company	Connections ('000)			Market share (%)		
	1992	1998	2000	1992	1998	2000
Residential connections						
BT	19 279	20 130	19 940	99.23	85.98	80.3
Kingston	130	157	177	0.65	0.67	0.71
Cable & Wireless	neg	953	n.a.	neg	4.07	–
Cable	23	2 173	4 714	0.12	9.28	19.9
Total	19 432	23 412	24 830	100	100	100
Business connections						
BT	5 859	7 713	8 689	97.18	89.07	85.1
Kingston	32	44	56	0.53	0.51	0.55
Cable & Wireless	132	458	417	2.19	5.29	4.08
Cable	6	291	819	0.10	3.36	8.01
Others	0	153	231	0.00	1.77	2.62
Total	6 029	8 659	10 212	100	100	100

Sources: Gist, (1999); Oftel market statistics

This is, however, in advance of the position in other western European countries as can be seen by looking at Table 2.2 which presents a comparison between the UK, Germany, Italy, Sweden and France on their incumbent's market share in 1998–99 to demonstrate how competition is gaining ground in the UK market compared to others in the region. It appears that the UK is performing well to prevent its incumbent (BT) increasing its market share in recent years. The BT market share for all reported services has been lower except for the German incumbent in national call service, although it should be noted that liberalisation in these countries started later than in the UK and has arguably moved at a faster pace.

The position in the mobile market is different, as it has developed from a duopoly to an oligopoly, as illustrated in Table 2.3. As we can see these market shares have been in the context of a major increase in the number of subscribers since the end of 1998, brought about in part by the development of 'pay as you go' mobile phones, which have been tremendously popular. Vodafone has been a major success with a market share of over 30 per cent by volume but a higher share by value in the UK. In addition, there is also evidence that Vodafone's return on sales and capital employed are higher than its competitors in the mobile market, which include BT Cellnet (Oftel February 2001a).

Table 2.2 Market share of incumbent operators in selected EU nations, 1998–99 (%)

	Local calls	National calls	International calls	Mobile
UK	80	70	55	30
Germany	95	65	65	38
Italy	98	94	85	57
Sweden	97	70	63	45
France	96	96	96	46

Source: Oftel (May 2000b)

Table 2.3 Mobile market share (%)

	1992	1995	1998	1999	2000
BT Cellnet	43	44	34	29	22
Vodafone	57	46	38	34	37
One2One	–	7	13	20	17
Orange	–	3	15	17	24

Sources: Oftel (May 2000a); MMC (1999) Table 4.13; Oftel (May 2001), Table 3, Quarter 3, 2000–01

While all telecommunications markets are growing, the growth in internet and mobile markets is phenomenal. Over the span of a year the growth rate in internet services was about 180 per cent, although from a very low base, and in 1999 and 2000 for mobile phones it was 84 and 67 per cent respectively. This creates an additional problem for the regulator with converging markets, which is discussed briefly later.

While it is clear from Table 2.4 that the internet market is the growth leader in the industry, the fixed line market, however, has arrived to a point of almost universal use as only about 5 per cent of UK consumers live in homes without a fixed line phone connection. Moreover, the strong demand in the internet market has expanded the choice of internet service providers (ISPs). Currently, over 400 ISPs are in operation in the UK. This huge number of ISPs has contributed in offering to the clients various services free of charge and unlimited usage packages. With all these developments some 37 per cent of UK households have access to the internet and just over 50 per cent of the adult population, which puts penetration above most of the European countries, with the exception of Scandinavia and the Netherlands and below penetration levels in the United States (Oftel July 2001c, para. 2.4).

Table 2.4 Growth in telecoms market, 1998–2000

	Market	Number of subscribers	% increase
March 1998–March 1999	Fixed Line	32.8 million lines	3
December 1998–January 2000	Mobile	23.9 million	84
January 2000–December 2000	Mobile	40.0 million	67
December 1998–January 2000	Internet	7.0 million	180

Sources: Oftel (May 2000a); Oftel Market Information Mobile Update (May 2001) Table 9

The mobile market apparently generates more revenue than other services per capita, due to the nature of products supplied and their business characteristics. A study in 2000 by Oftel and the Consumers Association suggests that there is some limited call substitution taking place between the fixed line phones and mobile users. About 25 per cent of mobile users reported that they use their fixed line phone less as a result of using their mobile and there is also some evidence of residential consumers giving up their fixed lines and relying exclusively on mobile phones. This has major revenue implications to these markets in the future.

The overall performance of the industry in recent years in revenue terms has also progressed steadily. According to Oftel (May 2000a), growth in the use of telecommunications services resulted in total turnover of £27 billion in 1999–2000, a growth rate of about 13.4 per cent from the previous year. However, recently scepticism has been expressed about the future profitability of the European telecommunications industry as a whole, largely because of the high prices paid by the operators for the third-generation mobile phone licences and the resultant large debt burden placed on the companies. This has been, certainly in the short-term, a major problem for the telecommunications industry and it may have longer-term repercussions. Space does not permit a proper analysis but the episode is worthy of study by those planning to auction licence rights.

4.2 Price Control

Since privatisation in 1984, the price control issue as one of the key areas of regulation has been debated extensively in the UK. Before the privatisation of

BT, a study by Professor Littlechild (1983) on BT's profitability devised a method of price regulation called RPI–X price cap regulation. This approach has been adopted to regulate all the privatised utilities in the UK and has also been adopted in Australia, New Zealand and other parts of the world.

What are the objectives of this method? Littlechild listed five criteria:

- To protect the consumers against monopoly
- To encourage the firm to achieve further efficiency gain and innovation
- To minimise the burden of regulation by having a pre-determined time frame
- To promote competition
- To proceed from privatisation to the next steps for the firm (Armstrong et al. 1995, p. 167).

The central part was a provision in BT's licence that prices should not increase by more than the retail price index[6] minus X percentage points, with X being a specified number.[7] This formula controlled a group of prices, which were known as the 'tariff basket' and were subject to an overall price cap. How this works is that the company is only allowed to raise the prices in this basket on average by RPI–X. Within the basket, however, the relative prices of these services can change. Thus, for example, within its basket, BT after privatisation relatively lowered the prices for long distance calls and raised the prices for local calls, all the time remaining within the price restraint. This freedom to 'rebalance' potentially has distributional consequences, as some groups of customers can be better off than others by virtue of this rebalancing. Thus a critical issue is exactly what services are included within the cap; a broad definition gives the company concerned more freedom to re-balance than a narrow one.

The point of capping prices rather than profits is that, unlike US rate-of-return regulation, it provides an incentive for a company to improve its efficiency. The X in the formula is premised on the company having a particular level of costs. If it reduces its cost base below this during the currency of the formula, it is able to keep the excess in the form of profit. This creates one problem because a way for a company to reduce its costs is to reduce the quality of the product offered. Thus it would be sensible to give the regulator powers to ensure the quality of service is maintained, but, as discussed above, this was not done in the 1984 legislation.

4.2.1 How does the price cap work?

An example is provided in Table 2.5 taken from an Oftel report in 1998. In the year 1997–98, BT was required to reduce the prices by RPI–4.5 per cent. This reduction accounted for £44.3 million in revenue terms. Table 2.5

demonstrates the cumulative and step-by-step reduction in prices by BT over 1997–98.

Table 2.5 Progress towards RPI–4.5% requirement 1997–98 (in million £)

1 October 1997	1 January 1998	17 January 1998	10 March 1998	29 April 1998
11.8	12.3	19.9	33.3	45

Source: Oftel (1998)

Prices of the following BT products were reduced to achieve the goal set by the RPI–4.5 per cent mechanism:

- 10 per cent reduction on national evening and night calls
- reduction on all calls to Japan and on weekend calls to France, Germany, Australia, and New Zealand among others
- 10 per cent reduction on national and regional weekends calls
- the discount received by selected Option 15 customers
- 10 per cent reduction in the price of local evening calls

On top of these price reductions to satisfy the regulator's ruling for 1997–98, BT has also been reducing prices in other areas, for example, a discount tariff to new businesses (Oftel, 1998).

Price control was a very controversial subject from 1984 up to around 1997 (Prosser 1997, pp. 66–71), in part because of a perception that the government had set the original price control in a way that was favourable to BT and also because of BT's re-balancing of its prices by lowering the price of long-distance services and raising the cost of residential services. Oftel's general response, in the price reviews of 1989 and 1991, was to tighten the price control by raising the value of X, from its original 3 to 7.5 by 1993, and to widen the coverage of the tariff basket. The effect of this was that the amount of the turnover covered by the price control rose from about half of BT's turnover originally to around two-thirds by around 1995.

Oftel then undertook a fundamental review of price control and its final proposal was that there should be a price cap covering those services which were already controlled but applied to the residential market only. In addition, the price cap would be applied using only the revenues and calling patterns of the first 80 per cent of residential customers by bill size. The value of X was to be 4.5, which Oftel argued was equivalent to 7 to 8.5 on the old basis, as BT had less scope for re-balancing. To protect small businesses, Oftel required BT to offer a controlled package, which was broadly similar to the residential control. Aside from this, there were to be no price controls in the business market. In addition, Oftel also inserted a new licence condition into BT's licence (the fair trading condition), and subsequently other operator's licences, prohibiting the abuse of a dominant position and agreements that distorted competition, modelled on Articles 81 and 82. Once this was in place, many of the detailed conditions in BT's licence were deleted. Finally, as regards interconnection charging arrangements, an annual determination of charges is in place with a system of network controls for those services, which were not competitive, using baskets of interconnections services, each subject to an RPI–X cap. So the outcome has been a business market regulated simply by competition law mechanisms and a domestic market where BT, although still regulated, is in theory subject to nation-wide competition, although Oftel extended the price control in this area until July 2002, with the possibility of a further extension to 2003 (Oftel February 2001b).

4.3 Trends in Prices

In this part of the study our primary aim is to provide the reader with an overall picture of price movement of various telecoms services in the UK between 1984 and 2000. Also, we are going to make comparisons of the UK price movements with that of the movements in other selected member countries of the EU. According to a recent study by Oftel (May 2000a), between 1984 and 1999 prices for main fixed-link BT services (final consumption) have fallen in real terms by 50 per cent. Prices have been dropping consistently about 1 per cent each year over recent years. These price drops were possible because of BT's compliance with the regulator's pricing policies between 1984 and 1997. A summary of the different levels of X for BT's retail price control for the entire period is presented in Table 2.6.

Table 2.6 BT retail price control, 1984–2001

	1984–89	1989–91	1993–97	1997–2001
X in RPI–X formula (%)	RPI–3	RPI–4.5	RPI–7.5	RPI–4.5

Sources: Armstrong et al. (1995); Gist (1999)

Table 2.7 summarises the changes in BT's regulated prices between 1984 and 1997 for selected services. The figures show an overall reduction in price levels for the services selected – line rentals would show a different picture. However, the reduction was substantial in the case of long-distance and international calls where the promotion of competition has been found more successful.

Table 2.7 Changes in BT's regulated prices of selected final services, 1984–97

Changes in price of:	Annual average % change	
	Nominal	Real[a]
Local calls (Standard rate)	+0.4	–4.3
National 'a' calls (Standard rate)	–0.4	–5.1
National 'b' low-cost routes (Standard rate)	–7.3	–12.0
International calls[b]	–8.8	–13.5

Notes:
[a] Annual average change in the RPI between 1984 and 1997 was +4.7 per cent
[b] Nominal and real changes in international call prices for 1991–97 only

Source: Gist (1999)

A comparison of prices between selected member nations in the EU is presented in Table 2.8. The benchmarking of prices reported here covers only three markets: public switched telephone network (PSTN) for residence, PSTN for business and mobile and four nations: the UK, France, Germany and Sweden. Prices are based on a snapshot as at 29 February 2000. The results for the price index for the three services, which are expressed relative to the UK price-level, show a better position for the UK in mobile phone service only, where the UK has on average the lowest prices against the countries considered. In the case of PSTN business service, the UK price is 27 per cent dearer than Sweden. For PSTN residential consumers, the price level in the UK is on average 8 per cent higher than the cheapest country, once again, Sweden. The UK also trails France by having 3 per cent higher prices in the business market. While it is hard to be certain with such selective results, the UK's mobile market seems more competitive than those of the other three countries. As shown in Table 2.3, in the UK's mobile market, four firms have been competing intensely to capture new market share, thus making the mobile end of the industry more competitive. Although it is necessary to be very cautious in making these comparisons, because they are sensitive to changes in methodology and therefore only the biggest differences are significant, possible reasons for the difference in the fixed network could be that in the UK it was less competitive; or that UK regulation was less effective than in the other three countries.

Table 2.8 Price indices of the three member nations in the EU, February 2000

Country	PSTN residential	PSTN business	Mobile
France	114	97	201
Germany	106	103	137
Sweden	92	73	113

Note: UK prices are regarded as 100

Source: Oftel (May 2000b)

4.4 Financial Performance of BT

Because BT has been the dominant operator throughout this period we examine its overall financial performance here. The performance of BT over 1981 and 1984 is presented in Table 2.9. It appears that the telecommunications sector has been a growth industry since 1981. The investment on tangible fixed assets has only declined as a percentage of turnover over 1981 and 1984.

Table 2.9 BT's performance, 1981 and 1984

Item	1981	1984
Turnover (£ million)	4 554	6 830
Operating profit (£ million)	753	1 531
Return on capital employed (%)[a]	9.4	16.7
Return on sales (%)	16.5	22.4
Investment as percentage of turnover (%)[b]	29.0	22.0

Notes:
[a] Capital employed is defined as total asset less current liabilities
[b] Investment is defined as the purchase of tangible fixed assets

Source: Armstrong et al. (1995)

Table 2.10 provides BT's performance between both the period 1985 and 1991, when BT operated as the incumbent privatised entity and Mercury as the second private carrier, as well as figures up until 1998, provided for comparison. Industry growth was phenomenal during this period and it appears that the growth rate during the duopoly period surpassed the growth in the pre-privatisation period, if measured by growth in turnover.

Table 2.10 BT's performance in duopoly period, 1985–98

Item	1985	1991	1993	1998
Turnover (£ million)	7 585	13 154	13 242	21 081
Operating profit (£ million)	1 856	3 531	1 927	3 341
Return on capital employed (%)	18.3	22.4	14.6	17.9
Return on sales (%)	24.5	26.8	18.5	16.0
Investment as percentage of turnover	24.0	22.0	–	–

Source: Armstrong et al. (1995)

BT's performance continued to improve between 1993 and 1998 as it made progress in both revenue and profitability terms in an absolute sense. Although it appears that the return on sales is about 1.5 per cent less in 1998 than in 1993, the rate of return on capital is higher for 1998 by more than 3 per cent. This indicates a better performance on the part of BT in 1998 over 1993.

However, 1998 arguably represents the peak of BT's performance, as its financial performance since then has been disappointing. We can illustrate this through examining the proportion of BT's pre-tax operating profits to its turnover, which is given in Table 2.11. Starting with 1984, the five-year average (unweighted) of the proportion was around 20 per cent until 1998. It then deteriorates steeply over the next three years and actually becomes negative in 2001. The three years from 1999 inclusive produce a proportion of 9.1 per cent, which goes up to 13.8 per cent if 1997 is taken as a baseline. Explanations for this disappointing performance are controversial. BT itself seems to lay the blame on increasing competition and the price controls imposed on its fixed-line and, since 2000, its mobile phone business. Financial reporters have laid the blame on BT's management, accusing them of failing to develop new ventures and of having concentrated too much on the old fixed-line business. In addition, BT has become saddled with a heavy debt burden as a result of its successful participation in the auction for third-generation mobile phone licences, which has also depressed its financial performance. As a result, a new Chairman has been appointed for BT, Sir Christopher Bland, and relatively radical new plans have been unveiled for splitting the group into its constituent parts; BT Wireless and BT Future, as well as selling off the Yellow Pages business (BT Yell). Indeed, BT has had approaches from one US group offering around £8 billion for its local loop business and a German group offering £18 billion for its national fixed-line network, both of which it rejected (*Financial Times*, 5 August 2001).

Table 2.11 Proportion of BT's pre-tax operating profit to its turnover

Year	Pre-tax profits (£m)	Turnover (£m)	Proportion of pre-tax profits to turnover	Five year average
1984	990	6 876	14.4	
1985	1 480	7 653	19.3	
1986	1 810	8 387	21.6	
1987	2 067	9 339	22.1	
1988	2 292	10 185	22.5	20.0
1989	2 437	11 071	22.0	
1990	2 302	12 315	18.7	
1991	3 075	13 154	23.4	
1992	3 073	13 337	23.0	
1993	1 972	13 242	14.9	20.4
1994	2 756	13 675	20.2	
1995	2 662	13 893	19.2	
1996	3 019	14 446	20.9	
1997	3 203	14 935	21.4	
1998	3 214	16 039	20.0	20.3
1999	3 188	*	18 223	17.5
2000	2 942	21 903	13.4	
2001	(1 031)	29 666	(3.5)	9.1 (from 1999)

Note: * This excludes as an exceptional item, BT's share of the proceeds of its sale of MCI shares

Source: BT Accounts

5. RECENT REGULATORY ISSUES

Of increasing importance as far as the United Kingdom is concerned has been the development of European Union policy towards telecommunications (Larouche 2000; Suater 1997, Ch. 5). Very roughly, this can be seen as having two strands: the development of competitive telecommunications markets throughout Europe and a concern with universal service. Insofar as developing competitive markets is concerned, the United Kingdom has been in the forefront of this, so that European Union policy has caused few problems, although there have had to be some changes to licensing procedures (Hunt 1997). The one important exception to this point is local loop unbundling, which is dealt with below.

5.1 Social Obligations Policies

Oftel identified universal service as an issue before the developments at EU level but they are now moving in tandem, as an EU Directive has specified that member states must make provision for universal service. The policy developed over a period of about four years, through a variety of documents from the European institutions, culminating in an amendment to the Voice Telephony Directive which incorporated the EC's up to date thinking on universal service.[8] The amended Directive defines universal service as:

> ... a defined minimum set of services of specified quality, which is available to all users independent of their geographical location and, in the light of specific national conditions, at an affordable price. (Article 2.2 (f))

Member states are directed to maintain, in particular, the affordability of services for users in high cost or rural areas, those with disabilities and those with 'special social needs'. This last phrase is critical because, although it is not defined in the Directive it has been taken, by the Department of Trade and Industry, to refer to low-income consumers (DTI 1998). This point gains further emphasis in Article 8 which says that member states shall, where appropriate, take specific measures to ensure equal access and affordability of fixed public telephone services for disabled users and users with special social needs. The Directive allows member states to set up financing schemes for certain services: the provision of network connections and telephone services, directory enquiries and public pay phones. The Directive also requires a range of consumer and quality of service issues to be addressed by the telecommunications operators, ranging from publication of information, to performance targets to rules on non-payment of bills. Equally important is the recognition that this is not a static issue, as the Commission is obliged to report on the functioning of the Directive, as well as having committed itself to providing monitoring reports on the provision of universal service across the Community (Commission of the European Communities 1999).

The debate in the EC has been paralleled by a similar debate in the UK, initiated by Oftel in the course of its review of BT's licence condition from about 1995. This review culminated in wide-ranging changes to BT's licence conditions, such as the introduction of a Fair Trading provision, as part of a move towards greater liberalisation. As part of this fundamental reconstitution of the telecommunications industry, Oftel engaged in serious development of the notion of universal service (Oftel 1995, 1997b, 1999 and September 2000). This exercise came to similar conclusions regarding the concept of universal service as the EC,[9] which has meant that it has been very easy for the UK to take on the obligations in the Revised Voice Telephony Directive.

These developments suggest a number of points. First, seemingly liberalisation cannot be accomplished successfully without paying explicit attention to issues of social obligations and universal service. Given the nature of a liberalised, competitive market, and that some of the obligation come from the EC, this means that they will have to be encapsulated in strict legal form. Admittedly, so far this has only been in a permissive and discretionary form, empowering the member states to do certain things and requiring the Director General of Telecommunications and the Secretary of State to take certain matters into account but their further elucidation may come in something akin to, if not identical with, a legally enforceable right.[10]

5.2 Local Loop Unbundling

There has been much controversy over the question of local loop unbundling.[11] The issue here is the ability of competitors to obtain access to BT's local exchanges so that they can install their own Digital Subscriber Line (DSL), which will enable them to provide high-speed internet access, and television pictures over the local loop. This would be in competition with BT which is currently installing its own version of DSL and thus has no incentive to be helpful. Oftel has been accused of acting with insufficient speed here by BT's competitors and there were reports that a legal challenge on the basis of the alleged dilatory nature of Oftel's action was being considered. The matter was complicated by negotiations over a EU regulation dealing with this same issue with a timetable for unbundling, in the initial draft, of January 2001 compared to Oftel's target date of July 2001. However, the issue was resolved by amendments to the regulation.[12] Oftel and the government have vigorously defended themselves against the charges, pointing out in their defence that the previous policy involved competing infrastructures, that is, encouraging the cable companies to provide an alternative source of supply to the local loop and that they have been moving as fast as practicable. However, by July 2001, it was reported that BT had only handed over 163 lines to competitors which either reflected a lack of demand from competitors or the slowness of BT's response, depending on whom you chose to believe (*Financial Times*, 26 July 2001).

5.3 Convergence

One particularly difficult issue has been that of convergence and the place of telecommunications in the context of utility regulation overall. The convergence issue arises because traditionally the telephone and broadcasting markets have been seen as separate, the former dealing in two-way voice and data communications, the latter in one-way broadcasts of sound and images and, in the UK, they have been subject to separate regulatory regimes. The last ten to fifteen years have seen the rise of a third market, mass computer networks that allow individuals to communicate, not usually using voice, and to have access to large amounts of information: text, images or sound. The distinctions between these markets are breaking down because of the increasing use of digital technology, which opens up the possibility of, for example, interactive links between a broadcaster and a viewer or the provision of a variety of services through the telephone system, such as video on demand.

The overriding question here is what is the appropriate regulatory structure for these new industries but this has become entangled with the more prosaic issue of regulatory reform. The Labour government, which came to power in the UK in 1997, promised to undertake reform of the regulatory system for utilities, which was taken to include telecommunications. The key elements of this reform were to be a revised set of duties for the regulators, the replacement of individual Director Generals with multi-person commissions, the creation of independent consumer councils, giving regulators power to impose monetary penalties for breach of licence conditions, regularising the provisions for altering standard conditions of licences and increasing the power of the Competition Commission in relation to licence amendments. The consultative process began in 1998 and culminated with the introduction of a Bill into the House of Commons early in 2000. The telecommunications companies, apparently not including BT, then lobbied vigorously against their inclusion in the Bill, on the grounds that they were not utilities in the sense of the energy and water industries and that, given the difficulties raised by convergence, it would be better to deal with all the issues at once.

The government acceded to this lobbying and removed the telecommunications provisions from the Bill, as well as the water ones, which had been the subject of a separate lobbying campaign. The result is that a number of these reforms, which would have dealt with some of the legal weaknesses identified above, have been put on hold. The government issued a White Paper on communications regulation (DTI 2000), which proposed the creation of a communications regulator in 2004, called Ofcom, which covers telecommunications, broadcasting and the internet and is responsible for both economic and content regulation. This means, however, that until at the earliest, 2003, the basic regulatory structure will remain that of the 1984 Act.

6. WHAT LESSONS FOR THE DEVELOPING ASIA–PACIFIC NATIONS?

Lessons from the UK experience are numerous. However, our aim here is to identify those lessons that are more appropriate for the Asia–Pacific environment than the environment that exists in Europe. As one study puts it,

'The experience of the UK is particularly relevant to the subject of telecommunications: how to regulate a liberalised market. We have what is probably the world's most liberated market, and I think I can safely say are more advanced than any other nation in Europe in this respect' (Butler 1998, p. 87). The Asia–Pacific developing countries, as we can see from the country study chapters in this book, are currently in the middle of liberalising their telecommunications markets and are at different stages of liberalisation. Some are at a transition stage, such as Malaysia, Sri Lanka and the Philippines, while others (Indonesia, Thailand, India, China and Vietnam) are still at a preliminary stage but have been taking genuine steps to liberalise the telecommunications sector. Under this circumstance, it will be hard to devise a uniform list of lessons to emulate for such diverse nations.

Lessons from the UK can be drawn mainly from the following issues: regulatory design, market structure, and social and economic benefits.

The major innovation in terms of regulation was the creation of an independent regulatory agency for telecommunications, which has operated without government interference in its day-to-day functions. In addition, the licensing arrangements described above have made it quite difficult for the Director General of Telecommunications to change the terms of the regulatory 'bargain' precipitately. Although there have been changes, notably with the ending of the duopoly policy and differing approaches to price control after 1996, these have taken place with a reasonable amount of warning and following established consultative procedures. Insofar as we can tell, there has been little political interference in major regulatory decisions and this is one of the most important lessons. If an independent regulatory agency is created, it must be insulated from political interference and the areas within which the politicians can interfere must be delineated clearly. How this can be done most effectively will vary from country to country.

One area where there has been criticism of the regulatory arrangements has been in the vesting of so much power in a named individual: the Director General. The consensus in the UK has moved away from this towards the appointment of Boards and Commissions, seen notably in the Utilities Act 2000 and the Postal Services Act 2000 and will also be a feature of the new communications regulator, Ofcom. The hope is that such a move will detract from an excessive personalisation of decisions that has occurred when individual Director Generals were responsible for decisions. It is also hoped that the Board approach will lead to greater continuity and stability even if at the expense of slower decision-taking on occasion.

Of crucial importance to stability and predictability is the use of open procedures of consultation. In this respect, Oftel has been very open and innovative, when viewed against the general run of governmental agencies in the UK. It has consistently published consultation papers and experimented with a variety of procedural devices to obtain the views of industry and other interested parties in the development of telecommunications policy. This relates as well to the agency being truly independent, because the consultation must be seen to be meaningful, as opposed to being undermined by governmental intervention behind the scenes.

In relation to the agency's powers, there are a number of points that can be made. First, it is important to establish a clear relationship between the agency's sectoral powers and whatever competition law powers exist. It is a matter for debate as to whether competition law powers should be vested in a sectoral agency or in a generalist competition regulator. Second, a regulatory agency needs power over quality of service as well as prices, particularly if the prices are controlled rather than based on return, as under the US rate of return system. Third, the agency needs proper rule-making powers, particularly when it is dealing with a competitive market. At the moment, the Office of Telecommunications lacks powers, for example, to impose standard licence conditions. Fourth, a regulator needs to have the backing of proper sanctions, in particular the ability to impose financial penalties, again something Oftel lacks at the moment.

In terms of market structure, the UK has been characterised by a cautious approach to competition and an emphasis on infrastructure competition and these points are interrelated. It is arguable whether infrastructure competition is the appropriate way forward, indeed from the European Commission's point of view, it would have been preferable to place more emphasis on the unbundling of the local loop, which has proceeded very slowly in the UK and is blamed by some for the slow roll-out of high-speed services. It looks now as if the key issue is not infrastructure competition but the sharing of common infrastructure and the terms on which competitors can connect to the network. This perhaps implies that the operation of the networks needs to be separated from the supply of services, as has been done in the UK electricity and gas industries. There is of course no guarantee that the previously dominant incumbent will remain successful with the coming of competition, the recent history of BT being a cautionary tale here. However, other companies may well take advantage of the competitive opportunities offered, the clear example being Vodafone which has been a very successful company having been built up only since about 1983. It should be emphasised that priorities are likely to be radically different in a country where telephone penetration is low and, correspondingly, the preferred competitive arrangements will be different. Specifically, in such countries the telecommunications industry needs to be designed primarily with a view to increasing line penetration, especially in rural areas, while in countries with a high telephone penetration the typical focus will be on price reduction, service innovation and quality improvements.

Finally, if one looks at the record in relation to prices and quality of service in the UK, we see a position that is much improved from that existing under a monopolistic nationalised industry. Prices have been reduced in real terms and quality of service has improved. Access to ordinary telecommunications services no longer seems to be an issue, in part due to the growth of mobile phone telecommunications. Clearly it is possible that the benefits of competition may not be distributed equally and the situation must be monitored but there is no evidence, in relation to telecommunications, that the less well off have suffered unduly. With the coming of convergence and the growth in information technology markets, this area will have to be monitored carefully but, if subsidies are to be given, this should be a transparent process.

If one examines the performance closely since 1984, the system seems successful. The prices of telecommunications services have been reduced, the quality of service has improved and customers now have a choice of suppliers and services (National Audit Office 1996). The regulatory system itself has developed in a notably open way, as far as Oftel is concerned. Once the forces of competition are unleashed, however, they develop a momentum of their own. Not only is it impossible to reverse the process, there are demands for it to be taken further. This affects the role of the regulator. The original scheme of the legislation viewed Oftel as an economic regulator, primarily concerned with controlling the prices that BT charged and the procedures reflected this narrow focus on one company. With the development of a competitive telecommunications market, Oftel now acts much more as an umpire between competing parties, a sectoral competition authority and a maker of rules, albeit within a legislative framework which still looks to the past. It looks now as if the regulatory system in the United Kingdom is about to reach another important turning point in the face of two separate issues. First, the need to deal with the convergence issue through the creation of a communications regulator and, second, the question of whether or not there is a role for sectoral regulation as opposed to general competition law. This again points up the importance of institutional arrangements and the need to have a sensible means for reviewing the effectiveness, or otherwise, of regulatory systems.

ACKNOWLEDGEMENT

The authors would like to thank Nicholas Garthwaite and Claire Milne for their helpful and perceptive comments. All errors of fact, opinions etc are, of course, the responsibility of the authors.

NOTES

1. See *Norweb Plc v Dixon* (1995) 3 All ER 952 at 959; *Timeload Ltd v British Telecommunications* (1995) EMLR 459.
2. It is a criminal offence to provide utility services without a licence, subject to some exceptions.
3. The current licence can be found at: http://www.dti.gov.uk/cii/regulatory/telecomms/telecomms_licences.shtml
4. But see *re British Telecommunications* OJ (1982) L360/36.
5. Available on the web at: http://www.telco-cpi.org.uk/
6. The primary index of consumer prices in the UK.
7. The following description is simplified considerably from the practice, which included a number of subsidiary price caps and special rules, such as a cap on line rentals.
8. Directive 98/10/EC the Revised Voice Telephony Directive (26.2.98)
9. Although perhaps the Community came to similar conclusions as Oftel!
10. 'Universal service is gradually evolving into a contractually enforceable right to affordable and effective service' in Hancher 1996, p. 137.
11. See *Financial Times* 19, 20, 22, 25 September 2000, 3 and 4 October 2000. *The Economist* 30 September – 6 October 2000.
12. EU regulation on unbundled access to the local loop (5 December 2000).

ABBREVIATIONS

BT	British Telecommunication
CC	Competition Commission
DGT	Director General of Telecommunications
DSL	digital subscriber line
DTI	Department of Trade and Industry
EC	European Community
IMF	International Monetary Fund
ISPs	internet service providers
MMC	Monopolies and Mergers Commission
OFT	Office of Fair Trading
Oftel	Office of Telecommunications
PSTN	public switched telephone network
ROCE	return on capital employed
RPI	retail price index
TA	Telecommunications Act

REFERENCES

Armstrong, M., S. Cowan and J. Vickers (1995), *Regulatory Reform, Economic Analysis and British Experience*, Cambridge, MA and London: MIT Press.

Bailey, K. and G. Carney (1995), 'The strange case of billing disputes', *Utilities Law Review*, **6**, 60–1.

Beesley, M. (1981), *Liberalisation of the Use of British Telecommunications' Network*, London: HMSO.

Better Regulation Task Force (2001) *Economic Regulators*, London: HMSO.

Bishop, M. and J. Kay (1988), *Does Privatisation Work?*, London: London Business School.

Butler, J. (1998), 'Regulating telecoms: Lessons from the UK' in P. Vass (ed.), *Network Industries in Europe: Preparing for Competition*, International Series 6, London: Centre for the Study of Regulated Industries, pp. 87–96.

Commission of the European Communities (1999), Fifth Report on Implementation of the Telecommunications Regulatory Package, November, Brussels: European Commission.

DTI (Department of Trade and Industry), (1990), *Competition and Choice: Telecommunications Policy for the 1990s*, Cm 1303, London: HMSO.

DTI (1991), *Competition and Choice: Telecommunications Policy for the 1990s*, Cm 1461, London: HMSO.

DTI (1998), *A Fair Deal for Consumers: Modernising the Framework for Utility Regulation*, London: HMSO.

DTI (2000), *A New Future for Communications* Cm 5010, London: HMSO.

Foster, C. (1992), *Privatisation, Public Ownership and the Regulation of Natural Monopoly*, Oxford: Blackwell.

Gist, P. (1999), 'Telecommunications: Costs, Accounts and Interconnection', in M.E. Beesley (ed.), *Regulating Utilities: A New Era?*, London: Institute of Economic Affairs in association with the London Business School.

Hall, C., C. Scott and C. Hood (2000), *Telecommunications Regulation: Culture, chaos and interdependence inside the regulatory process*, London: Routledge.

Hancher, L. (1996), 'Utilities Policy and the European Union', in P. Vass (ed.), *Regulatory Review 1996*, London: Centre for the Study of Regulated Industries.

Hunt, A. (1997), 'Regulation of telecommunications: The developing EU. Regulatory framework and its impact in the United Kingdom', *European Public Law*, **3**, pp. 93–116.

Larouche, P. (2000), *Competition Law and Regulation in European Telecommunications*, Oxford, Hart Publishing.

Lawson, N. (1992), *The View from No. 11*, London: Bantam.

Levy, B. and P. Spiller (1994), 'The institutional foundations of regulatory commitment: A comparative analysis of telecommunications regulation', *Journal of Law, Economics and Organisation*, **10**, pp. 201–46.

Littlechild, S.C. (1983), *Regulation of British Telecommunications Profitability*, London: HMSO.

Meek, C. (1998) 'Privatisation doesn't necessarily equal competition: the UK experience', in M. Hossain and J. Malbon (eds) *Who Benefits from Privatisation?*, London and New York: Routledge, pp. 100–18.

MMC (Monopolies and Mergers Commission), (1989), *Chatlines*, London: HMSO.

MMC (1995), *Telephone Number Portability*, London: HMSO.

MMC (1999), *Mobile Phone Charges*, London: HMSO.

Moon, J., J. Richardson, P. and Smart (1986), The privatisation of British telecom: a case study of the external process of legislation, *European Journal of Political Research*, **14**, 339.

National Audit Office (1996), *The Work of the Directors General of Telecommunications, Gas Supply, Water Services and Electricity Supply*, HC 645, London: HMSO.

Newman, K. (1986), *The Selling of British Telecom*, London: Holt Rinehart and Winston.

Oftel (1995), *Universal Telecommunications Services*, London: Oftel.

Oftel (1997a), *Review of Utility Regulation: Submission by the Director General of Telecommunications*,

Oftel (1997b), *Universal Telecommunications Services*, London: Oftel.

Oftel (1998), *News*, issue 41, September, London: Office of Telecommunications.

Oftel (1999), *Universal Telecommunications Services*, London: Oftel.

Oftel (January, 2000), *'Competition Plus' Strategy*, London: Office of Telecommunications.

Oftel (May 2000a), *Towards Better Telecoms for Consumers – May 2000 Progress Report*, London: Office of Telecommunications.

Oftel (May 2000b), *International Benchmarking Study of Telecommunication Services: A report by Teligen*, London: Office of Telecommunications.

Oftel (June 2000), *Encouraging Self- and Co-Regulation in Telecoms to Benefit Consumers*, London: Oftel.

Oftel (September 2000), *Review of Universal Telecommunications Services*, London: Oftel.

Oftel (February 2001a) *Effective Competition Review: Mobile*, London: Oftel.

Oftel (February 2001b) *Proposals for Network Charge and Retail Price Controls from 2001*, London: Oftel.

Oftel (May 2001) *Market Information Mobile Update*, London: Oftel.

Oftel (July 2001a) *Implementation of a Telecommunications Ombudsman* Scheme, London: Oftel.

Oftel (July 2001b) *Competition in the Provision of Fixed Telephony Services Ombudsman Scheme*, London: Oftel.

Oftel (July 2001c) *Oftel's 2000/01 Effective Competition Review of Dial-up Internet Access*, London: Oftel.

Pitt, D. (1990), 'An essentially contestable organisation: British Telecom and the privatisation debate', in J.J. Richardson (ed.), *Privatisation and Deregulation in Canada and Britain*, Aldershot: Dartmouth.

Prosser, T. (1997), *Law and the Regulators*, Oxford: Clarendon Press.

Sauter, W. (1997), *Competition Law and Industrial Policy in the EU*, Oxford: Clarendon Press.

Thatcher, M. (1994), 'Regulatory reform in Britain and France: organisational structure and the extension of competition', *Journal of European Public Policy*, **1**, 441–54.

Vass, P. (1995), 'Consumer representation – integration or independence?', in P. Vass (ed.), *Regulatory Review 1995*, London: Centre for the Study of Regulated Industries.

3. Australian telecommunications reform: Chasing a kangaroo with a butterfly net

Allan Brown and Justin Malbon

1. FROM MONOPOLY TO COMPETITION

The guiding principle in Australian telecommunications policy in recent years has been to open the industry to competition. As is evident from the other chapters in this book, this process is taking place in the context of similar telecommunications reform in other developed countries and in developing countries, and at a time when the industry itself is expanding rapidly both in terms of size and the range of available services and products.

Until the early 1990s, telecommunications services in Australia were provided by three carriers, each with a statutory monopoly in its respective area of operation. In 1946 the Overseas Telecommunications Commission (OTC) was established as the monopoly provider of overseas telecommunications services. The Postmaster-General's Department, established at the federation of Australia in 1901, was broken up into separate statutory authorities in 1975, with the Australian Telecommunications Commission (Telecom) taking over the former department's monopoly provision of telecommunications services within Australia, as well as its regulatory functions. In 1981 a third publicly-owned carrier, Aussat, was established as the monopoly operator of the domestic commercial satellite system, which commenced its activities with the launch of the first Australian satellite in 1985 (Brown 1991). The three carriers had exclusive access to their respective market segments guaranteed in legislation and were not permitted to compete with each other.

Minor and tentative steps towards liberalisation were made in the early 1980s when limited competition and private involvement were introduced in areas that did not overlap with the basic telecommunications network, for example, the provision of terminal equipment, private switchboards and paging services. In 1988 the Australian Telecommunications Authority

(AUSTEL) was established as a separate, independent authority to take over the regulation of the industry from Telecom, which was the major carrier. AUSTEL was given responsibility for protecting the monopoly position of the three carriers, safeguarding consumers from unfair carrier practices, technical regulation and promoting industry efficiency. Also during the 1980s the public carriers were corporatised, with various initiatives introduced to strengthen their commercial focus. Government concern shifted from the day-to-day decisions of Telecom, OTC and Aussat to the monitoring of their performance, while the carriers were each required to set financial targets and provide strategic corporate plans to the minister responsible for communications.

A much more far-reaching package of reforms was introduced in 1991 and implemented the following year. The government decided to merge Telecom and OTC into one general carrier, Australian and Overseas Telecommunications Corporation. (The corporate name of the merged entity was changed to Telstra Corporation Limited in April 1993, and in July 1995 its domestic trading name was changed from Telecom to Telstra.) Aussat was privatised by a trade sale to Optus Communications (now Cable and Wireless Optus,[1]) and the satellite system provided the basis for a second general telecommunications carrier in competition with Telstra. Mobile licences were granted to each of Telstra and Optus, and a third to another private company, Vodafone. Telstra's monopoly on the supply, installation and maintenance of fixed-line telephones was removed, as were restrictions on the purchase and resale of telecommunications capacity by carrier service providers (CSPs) (which in the Australian context are firms that supply telecommunications services using network infrastructure owned by carriers).

These reforms had the effect of shifting the satellite system from the public to the private sector, introducing carrier competition into the industry, replacing the monopolies for domestic and international services with a duopoly, and establishing a triopoly for mobile services. The government guaranteed Optus and Vodafone that the new market structure would remain in place until 1997. In particular, it would not issue additional general carrier or mobile licences before July 1997. In return, the general carrier licence granted to Optus imposed requirements upon the company in relation to the construction of specified network infrastructure.

The second wave of liberalisation in Australian telecommunications took place in July 1997, and effectively introduced open competition to the industry. From that date there has been no regulatory restriction on the number of carrier licences issued, minimal restrictions on the type of technology used, and no infrastructure rollout requirements imposed upon new carriers. Carrier licences are now issued for a fee based only on the recovery of the administrative costs involved. There is no restriction imposed on the number of carriers or CSPs licensed to operate, and, importantly, all carriers are required to interconnect all requesting carriers and CSPs (see s. 4.2).

2. THE AUSTRALIAN TELECOMMUNICATIONS INDUSTRY

The number of competing carriers in the Australian market increased from three at the end of June 1997 to more than 60 by the end of December 2000. The telecommunications industry in Australia was estimated to generate around A$26.5 billion (US$13.25 billion) in revenues during the 1999–2000 financial year, which accounted for around 4.5 per cent of Australian Gross Domestic Product, or A$1,500 (US$750) per person. In 1999 the penetration of fixed-line telephones was approximately 97 per cent of households, 48 per cent of households owned a mobile phone, and around 30 per cent had access to the internet (Paul Budde Communication 1999).

Table 3.1 sets out, for the financial years 1993–94 to 1999–2000, estimated revenues of the telecommunications market in Australia (excluding equipment) and the relative market shares for Telstra, Optus and all other service providers.[2] It can be seen that the telecommunications market grew substantially during the six years covered by the table – by 92 per cent in nominal dollar terms (from A$13.8 to A$26.5 billion). Whilst Telstra's revenues rose continuously during the period, its share of the total market fell steadily – from 91.3 per cent in 1993–94 to 68.3 per cent in 1999–2000. The dollar value of revenues accruing to Optus increased each year, as did its market share – from 4.3 to 10.2 per cent. Similarly, total revenues and market share for 'all other' industry players increased significantly over the period. (Throughout the remainder of this chapter all dollar values are expressed in Australian dollars.)

Table 3.2 presents selected financial data for Telstra for the same six-year period. The company more than doubled its profit from $1.7 billion in 1993–94 to $3.7 billion in 1999–2000. Return on assets and return on equity both increased substantially during the period.

These data illustrate that telecommunications is a highly significant industry in Australia and is growing rapidly. In terms of revenues, Telstra is clearly the dominant company, although its share of the market is declining from the effects of increased competition. Since market liberalisation, Telstra has earned a return on equity of around 30 per cent, which compares very favourably with the average rate for Australian firms of around 11 per cent (Fels 2001).

Table 3.1 Australian telecommunications revenue, 1993–94 to 1999–2000

	1993–94	1994–95	1995–96	1996–97	1997–98	1998–99	1999–2000
Telstra (A$ billion)	12.6	14.2	15.1	15.7	17.1	17.8	18.1
– (percentage of total)	(91.3)	(87.6)	(83.0)	(81.3)	(77.7)	(73.6)	(68.3)
Optus (A$ billion)	0.6	1.0	1.4	1.7	2.2	2.3	2.7
– (percentage of total)	(4.3)	(6.2)	(7.7)	(8.8)	(10.0)	(9.5)	(10.2)
All other (A$ billion)	0.6	1.0	1.7	1.9	2.7	4.1	5.7
– (percentage of total)	(4.4)	(6.2)	(9.3)	(9.9)	(12.3)	(16.9)	(21.5)
Total market (A$ billion)	13.8	16.2	18.2	19.3	22.0	24.2	26.5

Note: 1999–2000 values are estimates

Source: Paul Budde Communication 1999

Table 3.2 Selected financial data for Telstra corporation, 1993–94 to 1999–2000

	1993–94	1994–95	1995–96	1996–97	1997–98	1998–99	1999–2000
Profit after tax (A$ billion)	1.7	1.8	2.3	1.6	3.0	3.5	3.7
Return on assets (%)	15.4	13.6	17.5	10.1	20.1	22.7	21.2
Return on equity (%)	15.7	15.6	18.9	12.7	28.7	29.9	33.7

Source: Telstra Annual Reports

2.1 Local Access

Telstra owns the fixed public switched telephone network (PSTN) throughout Australia, and is the only carrier with nation-wide local coverage.[3] In the words of the government's primary economic research body, the Productivity Commission, 'Telstra's ubiquitous copper local loop is still overwhelmingly the dominant customer access network in Australia and is likely to remain so for the foreseeable future' (Productivity Commission 2001, page 4.1). Optus has offered local telephony over its hybrid fibre coaxial (HFC) cable network since 1997, but is limited to the three largest cities, namely, Sydney, Melbourne and Brisbane.

In terms of nation-wide market shares of basic access services, Telstra accounted for around 95 per cent at June 2000, Optus around 4 per cent, and other commercial business district (CBD) networks the remaining 1 per cent. The Optus local network passes slightly more than 30 per cent of Australian households. With the number of directly connected Optus local telephony subscribers growing rapidly (albeit from a small base), there is considerable scope for its share of local access revenues to increase.

2.2 Local Telephony

For 1999–2000 Telstra is estimated to have accounted for around 88 per cent of local fixed telephony revenues, and Optus around 3 per cent from its own network. In addition to Optus, competition to Telstra comes from resellers of Telstra's local telephony services. For 1999–2000 these accounted for around 9 per cent of total local telephony revenues.

Under Australian telecommunications regulation all carriers are required to price local calls on a flat rate ('untimed') basis. Price competition in the local call market started to emerge in 1997 when Optus introduced 20 cent local calls for subscribers directly connected to its network, compared to Telstra's 25 cent local calls. From August 1999 several resellers started to offer local calls at cheaper rates than those of Telstra and Optus. Telstra has responded by introducing a 15 cent 'neighbourhood call' for numbers called within the local exchange service area, and Optus has provided an attractively priced bundled offering of local access, local calls, domestic long-distance calls and pay TV subscription.[4]

2.3 Long-Distance Telephony

This segment consists of domestic long distance (STD) and international (IDD) calls. Telstra is by far the largest player in both, accounting for 75 per cent of total STD revenues for 1999–2000, and 48 per cent of IDD revenues. Although Telstra's shares of long-distance revenues are high, they have fallen significantly in recent years, especially for IDD where the share of the post-1997 new entrants doubled between 1998 and 2000 from 17 to 34 per cent.

The prices of long-distance telephony have fallen substantially. Between 1995 and 1999 the average real (inflation-adjusted) per minute price of STD calls declined by 37 per cent, and of IDD calls by 40 per cent. These trends in market shares and call prices are expected to continue.

2.4 Mobile Phone Services

In December 2000 there were six mobile phone carriers in Australia operating seven networks. With the population of Australia at 19.3 million, this was arguably an excessive number of competitors for the size of the market. By comparison, the United States had seven mobile phone operators, and the United Kingdom and Germany four each. A company named OneTel launched its GSM mobile service in May 2000, but substantial losses forced it out of business in May 2001.

There is a high degree of vertical integration between fixed-line and mobile services, with Telstra (46 per cent) and Optus (33 per cent) having a combined share of the mobile market for June–December 2000 of almost 80 per cent. The other original mobile licensee, Vodafone, had a 19 per cent market share.

2.5 Internet Services

There are in excess of 650 internet service providers (ISPs) in Australia. Only a small number have a national presence, however, and in some regional and rural areas there is only a single provider. Accordingly, the retail internet market is highly concentrated, with the largest three ISPs in July 2000 accounting for 50 per cent of the total market. Telstra (trading as 'Big Pond') had 24 per cent, OzEmail 18 per cent and Optus 8 per cent.

There are two trends bringing about increased competition in internet services. These are *free access*, where ISPs obtain their revenues from advertising rather than subscriber charges, and *bundling* of internet and telephony services, especially by Telstra and Optus. The recent introduction of high-speed internet services, via HFC cable, satellite and wireless technologies, is likely to further intensify competition among ISPs.

2.6 The State of Competition

In its 2001 study of the Australian telecommunications industry, the Productivity Commission concluded, in summary, that the more competitive sub-markets included long-distance telephony, mobile phone services and internet services, while effective competition was less well developed in local access and local telephony services (Productivity Commission 2001, 4.1). Telstra's shares of the various sub-markets are summarised in Table 3.3.

As indicated in s. 2.5 above in relation to internet services, the geographic scope of competition in telecommunications is largely determined by population density. Geography and population influence the number, density and demands of customers, and the per capita costs of deploying networks. The relevant geographic market segments for telecommunications in Australia are CBD, metropolitan, regional, and rural and remote areas. In capital city CBDs and to a lesser extent metropolitan and densely-populated regional areas, there is a high density of high-value subscribers and relatively low per-subscriber infrastructure costs. These factors encourage both facilities-based and service-based competition among a relatively large number of alternative providers.

In contrast, the costs of infrastructure rollout in rural and remote markets, compared to the resulting revenue stream, are generally such that it is economic for only a single facilities-based provider to serve the market. Most of the new carriers and CSPs that have entered the Australian market since

1997 have focused their infrastructure investment on CBD and metropolitan areas. However, some have sought to develop niche markets in regional and rural areas. While the entry of the new regionally-focused carriers and CSPs has had a limited impact on competition to date, it may have a greater effect in future years as their networks are developed and they gain a larger subscriber base (Productivity Commission 2001, Chapter 15).

Table 3.3 Telstra's share of telecommunications sub-markets, 1999–2000

Sub-market	Telstra's share (%)
Local access	95
Local telephony	88
Domestic long distance	75
International	48
Mobile phone services	46
Internet services	24

Source: Productivity Commission 2001, Chapter 4

3. PRIVATISATION OF TELSTRA

Although private interests in some countries, notably the United States and the Philippines, have historically owned telecommunications corporations, until the 1980s in most countries they were operated as government business enterprises (GBEs). There is now, however, a strong international trend towards the privatisation of publicly owned telecommunications carriers. The current wave of major privatisations in the industry began in the United Kingdom with Cable and Wireless in 1981, and since then many countries have followed suit. Various reasons have been advanced for the privatisation of telecommunications, including the desire to foster improved efficiency, to facilitate increased competition, to raise funds for the privatised enterprises to upgrade and extend their networks, and to assist governments in the repayment of public debt.

One of the major economic and political debates in Australia during the 1990s concerned the ownership of its national telecommunications carrier. This debate culminated in the partial privatisation of Telstra. The Labor party, which was in power in the federal parliament for 13 years until the March 1996 elections, was mainly responsible for the reform measures outlined in s. 5.1, but has had a consistent policy to maintain Telstra in public ownership. The Liberal party, however, has long advocated the privatisation of Telstra, and took steps to that end immediately upon achieving government in 1996.

Because of the high value of Telstra relative to the size of the Australian capital market, and because the Liberal–National party coalition did not have control of the Senate (the upper house of parliament), the government decided to privatise Telstra in stages rather than all at once. One third of Telstra was privatised in the first tranche (T1) in 1997, and another one sixth in the second tranche (T2) in 1999. At mid-2001, 49.9 per cent of the shares in Telstra were owned by private shareholders, with the remaining 50.1 per cent held by the Commonwealth government.

T1 took the form of a public share issue, which was the largest share offer in Australian corporate history and one of the largest world-wide in 1997. Total foreign investment was limited to 35 per cent of the shares issued, while shareholdings by individual foreign investors was limited to 5 per cent. Conditions imposed on the sale of shares in Telstra by the government included retention of its universal service requirements (see s. 5.2).

In Australia, the Finance Department has responsibility for supervising privatisation of government assets. The Minister for Finance decided that the issue price for T1 was to be $3.40 per share, payable $2 on application in November 1997, with the second instalment of $1.40 payable in November 1998. In the event, applications received for T1 totalled more than four times the number of shares available, with gross proceeds from the sale amounting to $14.2 billion (Australian National Audit Office 1999).

By the time the government obtained legislative approval for the sale of further Telstra shares, the stockmarket price of the T1 instalment receipts had risen considerably. Consequently, the shares in T2 were priced at $7.40 – $4.50 on application in October 1999, with the balance of $2.90 payable in November 2000. As with T1, applications for T2 substantially exceeded the shares available. Reflecting the higher price placed on the shares, proceeds for the 16.6 per cent of Telstra equity sold in T2 grossed $16.5 billion, more than the amount raised by the sale of twice the number of shares in T1.

The sale of shares in a GBE at a price below their market value represents a transfer of wealth from the public sector to private investors. It also involves a redistribution of wealth in favour of purchasers of shares at the expense of taxpayers generally. Vendor governments have an incentive to underprice shares issued in public floats of privatised corporations. Because they will suffer political embarrassment if applications fall short of the available number of shares and/or if individual investors or financial institutions sustain capital losses, governments will desire that each privatisation be 'successful' (Vickers and Yarrow 1988). Unsuccessful outcomes will cause difficulties for future privatisations. Governments may also underprice shares to encourage applications from those who do not usually invest in shares ('mums and dads'). If corporations are to be subject to regulation following privatisation, as is invariably the case with telecommunications, there may also be a 'sovereign risk problem' whereby uncertainty concerning the effects of regulation upon their future profits may increase the premium demanded by potential shareholders (Quiggin 1995).

The partial privatisation of Telstra provides an interesting case study regarding the pricing of shares for public issue. The price at which the T1 shares subsequently traded on the stock exchange was substantially *above* the original issue price set by the government in 1997, while the trading price for T2 proved to be significantly *below* their 1999 issue price. In a comprehensive study of Australian privatisations, Walker and Walker (2000, p. 225) were highly critical of the decision making in relation to the T1 sale process, and found that it was the most poorly handled of all privatisations they examined: 'on the basis of the gross transfer of wealth from the public sector to the private sector, the standout winner must be the sale of the first tranche of shares in *Telstra*'. Similarly, the Australian Auditor-General concluded that the T1 issue was underpriced:

> With the advantage of hindsight, the size of the initial listing premium and ongoing strong secondary market trading performance indicate that the issue was not fully priced. At the close of the first day of trading, Telstra instalment receipts [T1] stood at a premium of 67 cents, or 34 per cent over the institutional first instalment price of $2.00, the highest initial premium of any Commonwealth public share offering. (Australian National Audit Office 1999, pp. 13–14)

In common with telecommunications shares around the world, the value of Telstra shares fell considerably during 2000 and the first half of 2001 as equity holding in the sector was rerated. Although at the end of June 2001 the market value of Telstra shares was a substantial 58 per cent above the T1 issue price, it was 27 per cent below the issue price for T2. Taken together, the aggregate market value of Telstra shares issued in both tranches was approximately 14 per cent in excess of the privatisation proceeds received by the government.

The experience with the partial privatisation of Telstra illustrates the difficulty for governments in setting the 'right' sale price for initial share offers. The Australian government set the price for the sale of the first one third of Telstra 'too low', and was criticised for hurting taxpayers by under-selling their asset. It then sold another one sixth of Telstra for an 'excessive' price, and was criticised for hurting the large number of first-time 'mum and dad' investors whom the government had actively encouraged to take up shares.

The Labor party in Australia sees the retention of majority public ownership necessary for an intended 'nation building' role for Telstra, especially in relation to the further development of telecommunications in the non-metropolitan areas. In contrast, it is the policy of the Liberal–National government to sell off the 50.1 per cent of Telstra that remains in public ownership.

It is difficult to draw clear lessons from the partial privatisation of Telstra. Proponents of privatisation might refer to the company's performance and argue that the increase in profit in recent years is a result of privatisation. Privatisation sceptics might refer to the same performance and respond that Telstra's profit was increasing before privatisation and that more recent results simply represent a continuation of that trend. They may also point out that the company remains under majority public ownership, and that improved profit results can hardly be attributed to private shareholding. Privatisation advocates might further argue that the profits would have been even higher if the company were fully privatised.

In fact, the change in profitability of a newly-privatised enterprise is an inadequate indicator of the economic, let alone social, consequences of privatisation. As already mentioned, one of the reasons advanced for the privatisation of GBEs is to improve economic efficiency. However, efficiency concerns prices charged to consumers (allocative efficiency) as well as the costs of providing goods and services (productive efficiency). To the extent that higher profits after privatisation result from the privatised firm exploiting its market power by increasing its prices, there will be a reduction in allocative efficiency with consequent adverse effects for consumers and the economy generally. This is particularly relevant in the telecommunications industry where a formerly government-owned carrier usually enjoys substantial market power. This suggests that telecommunications industries should be fully liberalised to level and open competition *before* government-owned carriers are privatised. Otherwise, privatisation may simply result in the replacement of a government (quasi) monopoly for a private (quasi) monopoly.

Advocates of privatisation often refer to the traditional poor performance of GBEs with regard to productive efficiency, and claim that privatisation is the obvious and only remedy for such poor performance. This line of argument, however, ignores widespread evidence of greatly improved productive efficiency by GBEs, including telecommunications carriers, which has been achieved while still in public ownership. This transformation has come about largely through corporatisation, which is the process of introducing private sector principles and practices to public sector organisations (Brown 1997).

The data set out in Tables 3.1 and 3.2 on Telstra's revenues and profits showed how the substantial changes to its behaviour and performance commenced well before privatisation.

4. REGULATION

A major decision made as part of the reform of Australian telecommunications in the 1990s was to leave the ownership and control of the customer access network (CAN, or 'local loop') in the hands of Telstra, while its competitors depended upon access to Telstra's network for origination and termination of calls.[5] This is particularly notable considering that such vertical integration was contrary to the recommendations of the Hilmer Report (Hilmer et al. 1993), which was prepared by a government-appointed committee to advise it on the implementation of the nation's programme of microeconomic reform. The decision of the Labor government to maintain the vertically-integrated status of Telstra when it opened telecommunications to competition created a *structural imbalance* in the industry. The government had a range of options available to it other than merging Telecom and OTC, which would have resulted in more even competition. However, it decided to preserve the predominant position of the publicly-owned carrier to secure support within the Labor party and the broader Labor movement for the liberalisation of telecommunications. This decision was described at the time as a 'necessary political compromise' (Maddock 1992, p. 256). The structure of Telstra was left unaltered by the Liberal–National government when it assumed power. The vertical integration of Telstra was thus intact when it was partly privatised in 1997 and 1999, and still remains in place.

Of course many other developed and developing countries have introduced competition in their telecommunications industry without structurally separating the incumbent dominant carrier. Nevertheless, there are indications of a move towards the break-up of full-service telecommunications carriers. This move is mainly in response to pressures from financial markets where institutional investors are expressing a preference for telecommunications companies with 'transparent asset values', which are not available with incumbent carriers that combine infrastructure and service operations (Lacy 2000).

The Australian Information Economy Advisory Council (AIEAC) addressed the vertical separation issue in its 1999 National Bandwidth Inquiry (AIEAC 1999). It claimed that the break-up of a modern, vertically-integrated telecommunications carrier would be both practically difficult and expensive. In particular, the removal of the economies of scale and scope available to an integrated incumbent would have an adverse effect on its costs. The AIEAC also argued that there is mounting evidence that the CAN is no longer a natural monopoly. Technological developments are rapidly changing the economics of local-loop competition, and a number of operators around the world are providing CAN facilities in competition with the incumbent carrier. Indeed, as indicated in s. 2.1, Optus and specialised CBD operators do this now to a limited extent in Australia.

The AIEAC conceded, however, that control of the CAN facilities by Telstra does constitute a source of market power. The primary objective of the Australian government's telecommunications policy for the past two decades has been to encourage and enhance competition in the face of the market power enjoyed by Telstra from its ownership of the CAN and other incumbent advantages. From 1990 to 1997 the main strategy of this policy was to allow Optus to gain sufficient size and financial strength by establishing its market base, developing expertise and building infrastructure prior to the introduction of open competition in 1997. In 1991 the then Minister announced that the government's aim was 'to ensure that by 1997 there is a significant, recognised network competitor to [Telstra]' and that 'the creation of one strong competitor is the best way of achieving genuine and sustainable competition quickly'. To achieve this goal it was necessary 'to assist the second carrier [Optus] in overcoming the formidable advantages associated with [Telstra's] control over the customer base, infrastructure and access to information so that the second carrier will be able to compete from as level a playing field as possible' (Beazley 1991, pp. 3096–7).[6]

Following the introduction of open competition in July 1997, the responsibility for telecommunications policy-making remained with the Department of Communications, Information Technology and the Arts (DCITA). Regulation of competition in the industry, however, was transferred from AUSTEL to the Australian Competition and Consumer Commission (ACCC). Spectrum planning and the administration of the universal service requirements were transferred from AUSTEL to another government regulatory body, the Australian Communications Authority (ACA). Many of the changes to regulation were contained in the Telecommunications Act 1997. Unlike most countries, the primary Australian telecommunications regulation is undertaken by a general-competition regulator, the ACCC, rather than a telecommunications-specific (or communications-specific) regulator.

Nevertheless, amendments made in 1997 to Australia's main general-competition legislation, the Trade Practices Act 1974 (TPA), contained telecommunications-specific provisions, namely Parts XIB and XIC. Part XIB is intended to act as a deterrent to anti-competitive conduct in the telecommunications industry, and also gives the ACCC special powers to obtain information from carriers. Part XIC contains a telecommunications-specific access regime (see s. 4.2).

Major areas of industry regulation relate to:

- accounting separation;
- interconnection;

- preselection and number portability; and
- price regulation.

4.1 Accounting Separation

Telecommunications companies have a high proportion of fixed to variable costs, and of joint costs. Also, as capital and additional capacity are usually acquired in large and indivisible lumps, incremental costs can be difficult to separate. The conceptual and practical difficulties of cost allocation in multi-product firms are thus accentuated in telecommunications. The problem this provides for regulation is to prevent the network owner cross-subsidising services facing competition by proceeds from those areas relatively sheltered from competition. Information asymmetry between the established carrier and the regulatory authority, in relation to capital and operating costs, can exacerbate this problem.

The Australian telecommunications legislation addresses this issue by requiring *accounting separation* by all carriers. Prior to 1997 AUSTEL developed a chart of accounts and cost allocation manual (COA-CAM) for the industry. The COA-CAM specifies the principles to be followed by carriers in allocating costs to various areas of their activities and in reporting financial data to the regulator. The primary intention of accounting separation is to require Telstra to engage in arm's length transactions between different parts of its operations, and in particular for its internal transfer prices to be based on cost. The COA-CAM provides that any transfer price used by Telstra should be the price it charges to its customers, and, where the service does not have a market, to be based on long-run incremental cost. Accounting separation also provides the regulator with information for use in other regulatory functions, especially the determination of interconnection charges and price regulation (see below).

4.2 Interconnection

The introduction of competition in telecommunications is usually achieved by providing new entrants access to the incumbent's local network. The technical and financial barriers facing new entrants are much greater in relation to the provision of the local segment of calls than for long-distance segments, and interconnection avoids the need for new entrants to bear the high cost of duplicating the incumbent's local network. A problem with this form of competition, however, is that there is a conflict of interest between the incumbent and new entrants over the *price* of interconnection.

Under Part XIC of the TPA the ACCC decides which services are within the scope of the access regime ('declared services'). Telstra's domestic PSTN originating and terminating access services have been 'declared' by the ACCC. This has two important consequences. First, Telstra must supply these services to each of its competitors upon request. Second, if a competitor cannot agree with Telstra about the terms and conditions of supply (for example, price) then one of them can notify the ACCC of an access dispute. The ACCC can then arbitrate to resolve the dispute. This includes setting the price on which Telstra must supply these services. Alternatively, rather than wait for the ACCC to arbitrate the dispute, Telstra can give the ACCC an undertaking setting out proposed charges for access. The ACCC can then accept or reject the undertaking.

Interconnection pricing is fundamentally important, and inherently complex and controversial. Because Part XIC is largely silent on the principles that should govern access pricing for telecommunications, the ACCC has significant discretion in determining the pricing approach for declared services. In common with several other developed countries with recently liberalised telecommunications markets, the ACCC strongly favours the 'total service long-run incremental cost' (TSLRIC) costing methodology to determine interconnection charges payable to access providers. This approach attempts to estimate the annual additional cost incurred by the access provider (Telstra) in the long run in providing interconnection to competing carriers and CSPs, assuming all of its other production activities remain unchanged. Alternatively, it can be viewed as the total annual cost that Telstra would avoid in the long run if it ceased to provide interconnection.

Since 1997 the ACCC has issued a number of guidelines to assist parties in its interpretation and application of the TSLRIC methodology. Nevertheless, by 2001 the intended purpose of the guidelines, namely to encourage agreement on access pricing between parties rather than approaching the ACCC for arbitration, had not been achieved. Nor had the prices in the few undertakings submitted between 1997 and 2000 to the ACCC by Telstra approximated the prices arrived at by the ACCC (Productivity Commission 2001, 10.4).

With the commencement of the interconnection requirements in July 1997, Telstra had introduced an initial interconnection charge of 4.7 cents per minute. Access seekers contested this rate and the matter was referred to the ACCC. Following a series of protracted hearings and negotiations, Telstra made an undertaking proposing the charge be reduced to around 3.5 cents per minute. In September 2000 the ACCC rejected Telstra's undertaking and determined that the charge should be reduced by a much greater proportion, to 1.53 cents per minute (which is around 33 per cent of the original charge, and 44 per cent of that proposed by Telstra).

The main point of contention between Telstra and the ACCC is that Telstra argues that the determination should be based on its *actual* cost of providing interconnection, whereas the ACCC adopts an *efficient* cost approach. In other words, in its application of the TSLRIC methodology, the ACCC estimates the cost to Telstra of providing access, not on its *existing* network, but on a *forward looking* network. The ACCC claims that an efficient access provider could provide the interconnection services at a lower price than that offered by Telstra:

> Telstra has lobbied strongly for access arrangements to be based on actual costs. In the ACCC's view, the use of actual costs for determining access prices would have a range of adverse consequences. Actual costs are likely to be inflated by monopoly profits or inefficient network and other work practices, which may be embedded in Telstra's existing cost base. Telstra should not be entitled to pass on its own inefficiencies to its competitors. (ACCC 2000)

This point of contention between Telstra and the ACCC remained unresolved at the end of 2002.

4.3 Preselection and a Number Portability

Competition in telecommunications is inhibited to the extent that subscribers are discouraged from or inconvenienced by switching between carriers. To foster competition a policy of 'equal access' has been adopted to enable telephone subscribers to be able to use the services of competing carriers as easily as those provided by Telstra.

When Optus commenced operations in 1992 subscribers needed to dial '1' before their required number in order to have their call carried by Optus. This was an interim arrangement until Telstra's network was upgraded to facilitate 'preselection'. Between 1993 and 1997 AUSTEL conducted of a series of ballots, on an area basis and starting with the major capital cities, whereby all Australian telephone subscribers were issued with ballot papers and invited to specify their choice between Telstra and Optus as their preferred long-distance carrier. Around 14 per cent of subscribers chose Optus and were transferred to that company for all STD and IDD calls, while the remaining 86 per cent who specified Telstra or did not participate in the ballot remained with the incumbent. Subscribers were able to override their preselected carrier (without charge) and have their call carried by the alternative carrier by dialling a four-digit code before the relevant long-distance number.

Preselection provides an important safeguard against carriers 'locking in' customers. Since 1997 preselection has been extended as a requirement upon all carriers: it is not only Telstra which must provide preselection, and subscribers can preselect any carrier. They can still override their preselected carrier for individual calls by dialling specific four-digit codes of alternative carriers. In addition, preselection now applies to calls that originate on fixed networks and terminate on mobile phone networks (fixed-to-mobile calls).

The ACCC and the ACA are also engaged on a programme of introducing 'number portability', which refers to the ability of subscribers to change their preselected carrier without having to change their telephone number. As with preselection, number portability is considered a crucial ingredient in promoting competition in a multi-service provider environment. It enables customers to make a choice between service providers based on price and service without the inconvenience and cost associated with changing their telephone numbers. Australian and international studies have clearly established that the absence of number portability substantially increases the reluctance of subscribers, particularly business subscribers, to change service providers (ACCC 1997). While incumbent firms can provide subscribers with number portability, they have little incentive to do so without regulatory intervention. The ACCC and ACA have thus intervened to make the practice mandatory. Number portability for fixed-line services was introduced in November 2000, and for mobile services in September 2001.

4.4 Price Regulation

The prices of telecommunications services in Australia have always been subject to some form of regulation. The regulation of telecommunications pricing was required when Telecom, OTC and Aussat were monopolies, and was still deemed necessary with the introduction of competition to prevent Telstra abusing its market power. Price regulation was exercised directly by the government, in the name of the Minister, until 1984 when it came under the purview of the price regulation body, the Prices Surveillance Authority. The regulation of telecommunications pricing passed to AUSTEL when it was established in 1989, and to the ACCC when it took over AUSTEL's functions in 1997.

Competition in telecommunications has necessitated a move towards 'rebalancing' Telstra's tariffs to bring them more closely into line with costs. For some services this process has also required 'de-averaging' which is a more specific form of rebalancing, and refers to charging prices related to the cost of providing individual services as distinct from the average cost of a similar cluster of services. Notably, the prices Telstra charged for STD calls were historically based on distance rather than the intensity of traffic, even though significant economies of scale were achieved on busier (thicker) routes.

Price caps were first applied in the United Kingdom in 1984 in relation to British Telecom (see Chapter 2), and have also been adopted in several countries for the price regulation of telecommunications as well as other utility services such as water, gas and electricity. A price cap involves two main variables, an inflation index – in Australia the Consumer Price Index (CPI) – and an 'X factor'. Under the price cap the average of a basket of prices is not permitted to increase in any period by a percentage in excess of CPI–X. The price cap, therefore, ensures that the average price of regulated services falls in real terms by at least the amount of the X factor. To maintain profit margins, firms subject to price-cap regulation must improve their overall productivity by at least the value of X, and any improvement in excess of X will, all other things being equal, increase their profitability.

Regulated firms thus face positive and negative incentives to improve productivity, and the resulting gains are shared with consumers by way of lower real prices. Price caps have been designed to apply to monopolies and other firms with substantial market power, and are intended to produce results – improved productive and allocative efficiency – normally associated with increased competition. An overall price cap allows a multi-product firm to move towards rebalancing its prices by increasing its prices for various products by differing proportions within the overall average limit. In the price-cap regulation for Australian telecommunications, the X-factor is set, not by the ACCC, but by the Minister. Nevertheless, price caps allow the government and the regulator to remain at arm's length from day-to-day involvement in detailed price setting.

In 2004 Telstra's prices (but not those of its competitors) are subject to price-control arrangements that have two major components. The first component is a price cap on a 'basket' of telecommunications services, including local call, line rental, mobile, STD and IDD services. The second component is a number of 'sub-caps' on particular services in the broad basket. These include a 22 cent price cap on local calls, a CPI–0 per cent price cap on a basket of local calls and line rentals, and a requirement for local call charges to be broadly the same for both metropolitan and non-metropolitan subscribers. There are also further sub-caps for 'low spend' subscribers.

Since its introduction there have been five distinct periods of price-cap regulation of Telstra:

- *July 1989 to June 1992: X = 4.0 per cent*
  The initial price cap covered the three-year period up to the introduction of limited carrier competition. A revenue weighting method was adopted to calculate an average allowable price movement for a basket of services consisting of rentals and local, STD and IDD calls.
- *July 1992 to December 1995: X = 5.5 per cent*
  For the second period, telephone connections, leased-line rentals and mobile phone services were added to the basket of services subject to the price cap, and the X-factor was increased to 5.5 per cent.

- *January 1996 to June 1999: X = 7.5 per cent*
  For the third period the price cap applied to a similar basket of services as for the second period, and the X-factor was again increased, this time to 7.5 per cent.
- *July 1999 to June 2001: X = 5.5 per cent*
  For the fourth period the X-factor reverted to the level that applied during the second period, namely, 5.5 per cent.
- *July 2001 to June 2005: X = 4.5 per cent*
  For the fifth period the X-factor was reduced further to 4.5 per cent.

Although the basic CPI–X formula might suggest an administratively simple form of regulation, in practice the regulator requires a substantial amount of information to assess the historic and prospective productivity of Telstra. This problem is exacerbated by the use of sub-caps in an attempt to restrain Telstra in its rebalancing. According to one commentator, the attempt to fine-tune the Australian telecommunications price-cap scheme 'has resulted in a complicated, more intrusive, less transparent system than was originally intended' (Xavier 1993, p. 42).

The price levels and trends of residential and business telecommunications services in Australia are regularly compared with those in a number of countries in the Organization for Economic Cooperation and Development (OECD). A study covering the period February 1998 to June 1999 compared certain telecommunications prices for nine countries, namely, Australia, Canada, Finland, France, Japan, New Zealand, Sweden, the United Kingdom and the United States. It found that since the introduction of open competition in 1997 telecommunications prices in Australia had fallen sharply, particularly for STD and IDD calls. Residential mobile phone prices in Australia were found to be relatively high compared with the other countries studied. However, for most services Australia was in about the middle of the benchmarked countries (Productivity Commission 1999).

In August 2000 the ACCC was directed by the Minister to undertake a review of the price-cap regime. The ACCC was required in its review to have regard to the current state of competition in each of the price-controlled markets, and the extent to which further efficiency gains may be achieved with and without the price-control arrangements. The ACCC was to advise on whether there was a need for the pricing restrictions on Telstra to continue after the expiration of the fourth price cap period in June 2001. If it considered that there did remain a need for price-control regulation, it was to determine what form such regulation should take. The ACCC completed its report in February 2001.

Its assessment of the state of competition in the various industry sub-markets was broadly consistent with that in the March 2001 draft report of the

Productivity Commission on telecommunications competition regulation. The ACCC concluded that most of the price-controlled telecommunications markets were still not subject to effective competition:

> the Commission believes competition has not yet developed fully in the markets in which local calls, line rentals and connection services are supplied. Further, the Commission is not convinced that competitive forces are sufficiently strong to achieve effectively competitive outcomes in the markets in which STD, IDD and fixed-to-mobile services are supplied. (ACCC 2001, p. xiv)

The ACCC also found that Telstra had kept its actual price reductions very close to the minimum required by the price-cap regulation. While the prices of Telstra's price-capped services had fallen substantially in real terms since 1989, the prices of services used by large-volume, high-expenditure business customers had generally fallen faster than prices of services used by residential customers. Similarly, volume discounts had tended to be focused more on business customers than residential customers. Moreover, the reductions in prices under the price caps tended to be concentrated in sub-markets subject to the highest degree of competition, namely, mobile and international calls. These findings are basically the same as those made by AUSTEL in its 1994 review of the price-control arrangements (AUSTEL 1994).

The ACCC concluded that the community as a whole would be better off by the retention of a single broad price cap on all price-controlled services. It recommended, however, that a number of existing sub-caps should be removed from the price-control arrangements to facilitate Telstra rebalancing its prices. The ACCC further suggested that the number of services included in the price-control arrangement be reduced over time:

> While price control arrangements can help protect consumers during the transition from monopoly to competition, there is also a danger they can inhibit the development of competition, efficient investment and the long term benefits of consumers. (ACCC 2001, p. viii)

By October 2001 there had been no response from the government to the ACCC's report apart from a decision to extend the fourth period price-cap arrangements from June 2001 to June 2002.

4.5 Productivity Commission Inquiry

As in most countries, the industry-specific regulation of telecommunications in Australia reflects the special nature of the industry and the inadequacy of general competition law as a regulatory tool. In particular, regulation of

Australian telecommunications implicitly recognises the considerable market power enjoyed by Telstra resulting from its vertically-integrated structure and its status as the incumbent carrier. Whereas general-competition regulation is mainly concerned with the *behaviour* of firms in the economy, at the current stage in the evolution of competition in Australian telecommunications, regulation is largely concerned to bring about a *structure* in the industry that is conducive to 'sustainable competition'. Although this concept has not been precisely defined by Australian policy-makers, it is related to the eventual attainment of a situation where competition is sufficiently robust for the industry to be able to be effectively regulated by general trade practices legislation, and for the telecommunications-specific provisions in the TPA to be repealed:

> It is intended that competition rules for telecommunications will eventually be aligned, to the fullest extent practicable, with general trade practice law. Part XIB will apply for the period from 1 July 1997 until some future review determines that competition is sufficiently established that the Part or some provisions of the Part are no longer needed. (Explanatory Memorandum, Trade Practices Amendment (Telecommunications) Bill 1996, cited in Productivity Commission 2001, 5.12)

In June 2000 the Productivity Commission was directed by the government to examine and report on the operation of Parts XIB and XIC of the TPA and certain provisions of the Telecommunications Act 1997. A key issue for the Commission was whether telecommunications should continue to be subject to telecommunications-specific legislation and regulation. In March 2001 the Commission released its extensive (500 plus pages) draft report (Productivity Commission 2001), which was the first comprehensive study of the Australian telecommunications industry since its liberalisation in 1997.

In its draft report the Productivity Commission recommended that the provisions in the TPA for the telecommunications-specific access regime (Part XIC) be retained. Nevertheless, it was critical of the existing access arrangements and procedures. While the regulation of access was intended to be relatively 'light-handed', in practice it had proved to be far from this. Essentially, the Commission concluded that the determination of access *prices* was the key to the access regime, but the current procedures of declaration, undertakings and arbitration were not well suited to this task. A number of reform options to address the problems with the access regime were proposed, which were to provide the basis for the Commission's recommendations in its final report. The Commission described the specific deficiencies in the access arrangements as follows:

> the declaration criteria in Part XIC are vague and provide excessive discretion to the regulator; the processes for determining conditions for access under Part XIC

> are slow and inefficient, reflecting the failure of undertakings as a mechanism and the predominance of lengthy bilateral arbitrations between conflicting parties; [there is] the risk of reduced investment in core telecommunications infrastructure ... [and] the lack of legislated pricing principles to guide access seekers, providers and regulators. (Productivity Commission 2001, p. xx)

In relation to Part XIB the Productivity Commission noted that it was intended to promote competition in telecommunications by facilitating speedy action against alleged anti-competitive conduct. It pointed out, however, that there had been only a few ACCC investigations under Part XIB, and where it had been used it had been complex to administer and action had been slow. The Productivity Commission considered that other mechanisms such as the general anti-competitive provisions of the TPA would be an effective alternative, and recommended that the telecommunications-specific provisions in Part XIB be repealed.

5. DELIVERING CONSUMER BENEFIT

As the delivery mechanism for information technology, telecommunications is recognised as an increasingly critical component of economic growth. National economic growth is becoming heavily reliant on the efficient and effective delivery of electronic information to drive down the information costs of businesses and consumers. For this reason telecommunications policy and regulation play a pivotal role in promoting economic growth and consumer benefit.

It was mentioned earlier that in 1993 the Hilmer Report was prepared to advise the government on the implementation of Australia's programme of microeconomic reform. The Hilmer Report recognised that the competitive-market mechanism cannot always be relied upon to promote the public benefit. Accordingly, processes have been put in place for dealing with consumer complaints against telecommunications service providers, and to address the issue of universal service.

5.1 Consumer Complaints

The overall credibility of the telecommunications industry relies on consumer confidence in the operations of the industry. Increased competition can have the adverse side effect of over-zealous and unethical service providers engaging in unfair and illegal practices to gain a market edge. In 1993 the federal government established the Telecommunications Industry Ombudsman (TIO) to resolve disputes between telecommunications companies and residential and small-business customers. The TIO is funded by the industry, and provides a free service to telecommunications customers. An independent board structure has been created in a way designed to protect the TIO from interference by the industry. The TIO has power to investigate complaints about billings, telephony and internet services, privacy and breaches of industry codes and standards. It can make binding determinations up to $10,000 against telecommunications companies and ISPs.

According to the TIO, the telecommunications industry is becoming increasingly complex, new products and services are constantly being introduced, and customer turnover ('churn') is increasing. Both industry and consumers are finding it difficult to keep up with the pace of change, which has provided scope for some unscrupulous practices to develop. In particular, there has been a sharp increase in consumer complaints to the TIO about 'slamming', a practice in which a competing phone company transfers a telephone service to that company without the subscriber's informed consent. Of the over 4000 complaints dealt with during 1999–2000, nearly 15 per cent related to slamming (TIO 2000).

5.2 Universal Service

Historically, one of the obstacles to private involvement and competition in telecommunications has been the requirement of governments upon incumbent carriers to provide certain services that are considered unprofitable because of geographic or other reasons. The provision of telephone services to the remote areas of Australia has long been a policy objective, together with the installation and maintenance of unprofitable public payphones. The remote-area requirement is claimed to be especially onerous on the basis that providing and maintaining services to customers in low-density rural and regional areas subjects the provider to high infrastructure and operating costs. The losses incurred in providing such unprofitable services were traditionally cross-subsidised from Telstra's profitable areas of operation. This became no longer appropriate with the introduction of competition, as it would have placed Telstra at an unreasonable competitive disadvantage.

The regulatory response to the provision of unprofitable telecommunications services was the introduction in 1991 of explicit requirements concerning the universal service obligation (USO) which are now administered by the ACA. The USO is defined by the Telecommunications (Consumer Protection and Service Standards) Act 1999 (TCPSS Act) as the requirement to ensure that standard telephone services and payphones are 'reasonably accessible to all people in Australia on an equitable basis, wherever they reside or carry on business'. The standard telephone service involves the provision of a voice grade telecommunications service, or the equivalent for people with disabilities, and incorporates aspects of social regulation such as continued access to untimed local calls. Telstra is currently the only declared universal service provider (USP) (but see below), and has the responsibility to provide service on request to anyone in the areas for which it is declared.

Since July 1999 the universal service regime has also comprised the digital data service obligation (DDSO). The main requirement of the DDSO, covering about 96 per cent of the Australian population, is that a data channel with a data transfer rate of 64 kilobits per second be available to all residents within the specified DDSO area on the same basis as Telstra's voice services. Currently Telstra is also the national DDSO provider (Productivity Commission 2001, 17.4).

The cost to Telstra of providing universal services is shared amongst all carriers according to their respective proportions of industry revenue. The costing mechanism involves the calculation each year of the net universal service cost (NUSC), which forms the basis for the universal service levy (USL) that is imposed on all carriers and payable by them to Telstra. There are a number of approaches to the financing of community service obligations, including direct government subsidisation. The levy mechanism used in Australia, however, places the cost of universal service provision on the telecommunications industry.

The methodology for calculating the annual NUSC is contentious and has changed several times since 1991. It is currently calculated by the '*avoidable costs* minus *revenue forgone*' formula. Avoidable costs are calculated on the basis of long-run incremental costs. In essence, avoidable costs are those incurred by a USP that it would not have incurred if it had not supplied services to high-cost areas. In order to set incentives for cost minimisation, the cost calculations are made on a forward-looking basis, that is, they are based on assumptions about technology that is deemed to be industry best practice rather than actual costs incurred. Revenue forgone is designed to measure the additional revenue that a USP collects from supplying universal services (Productivity Commission 2001, 17.5).

The methodology for calculating the NUSC is somewhat similar to TSLRIC costing employed for access pricing (section 4.2). Also in common with determining access costs, there are problems involved in the calculation of the NUSC. Not the least of these problems is that of information asymmetry, since the calculation of the NUSC requires the use of Telstra cost data. Telstra has a financial incentive to inflate the value of the NUSC so as to maximise the levy payments it receives from other carriers.

For the 1997–98 financial year, the first year for which the NUSC cost model was applied, Telstra's estimate of the NUSC was $1.827 billion. The government then intervened and introduced legislation to cap it at only

$0.253 billion, less than one seventh of Telstra's estimate. Although the ACA subsequently assessed it at $0.548 billion, the legislative cap prevailed. Telstra then suggested the NUSC for the subsequent two years, 1998–99 and 1999–2000, be based on the ACA's assessment of $0.548 billion. However, the NUSC for both 1998–99 and 1999–2000 was set by the Minister at $0.28 billion ($0.253 billion plus an adjustment for inflation) (Productivity Commission 2001, 17.4–17.5).

There are clearly major problems with universal service regulation in Australia. The purpose of the government effectively freezing the NUSC calculation process and the amount for the 1999 and 2000 financial years was to give it the opportunity to rethink its universal service policy. Amendments made to the TCPSS Act in 2000 establish a new experimental framework for the supply of universal services. This consists of (a) a three-year contestability pilot project, and (b) a competitive tender (the 'extended zones tender'). These arrangements, possibly the only ones of their kind in the world, are designed to introduce competition to the universal service process, to reduce the cost of universal service provision, and to simplify the task of regulation (DCITA 2000a). The implementation of these arrangements commenced in early 2001. The contestability pilot project is initially confined to two large regional areas, but if deemed successful is expected to be extended more widely.

In the pilot project, Telstra has been designated as the primary USP for each contestable area. As such Telstra is obliged to provide universal service to the two areas for the three-year period of the pilot project. There is also provision for one or more other carriers to be designated as competing USPs. Any competing USPs will be able to enter and exit the market at their discretion (with requirements for smooth transition). The universal service subsidy for contestable areas is to be given to primary and competing USPs equally on a per-service basis (Productivity Commission 2001, 17.6–17.7).

In October 2000 the government invited seven companies to 'tender' for the provision of calls in the 'extended zones', which are call charging zones that lie outside Telstra's standard local-call charging zones. The tender document required the provision of untimed local-call rates, as well as other specified services at capped prices. The successful tenderer is to be obliged to provide the tendered services for a period of ten years. It will have exclusive access to the universal service subsidy for three years, after which time the provision of services to these areas may become contestable. In February 2001 the government announced Telstra as the successful tenderer.

In its review of universal service, the Productivity Commission formed the opinion that longer-term competition among competing providers, as intended by the government, was unlikely to develop under these new arrangements unless the universal service levy was increased. It also recommended that the power to determine the levy should lie with the ACA, rather than the Minister (Productivity Commission 2001, 17.1).

The issue of universal access has gained added importance in relation to the desire of the Liberal–National coalition government to privatise the remaining 50.1 per cent public equity interest in Telstra. For the government to obtain political support to sell off the rest of Telstra, it must convince those living in the rural and regional areas that full privatisation will not threaten the availability, quality and cost of their telecommunications services. To address these concerns the government commissioned yet another inquiry into the telecommunications industry, this time into quality of services provided, especially in the non-metropolitan areas. The inquiry headed by Tim Besley reported its findings in September 2000. The Besley Report found that a significant proportion of those who live and work in rural and remote areas of Australia had unacceptably high levels of basic telephone fault rates, a lack of reasonable access to the internet over their telephone line, and that the data demand of some major organisations were not fully met by the telecommunications industry (DCITA 2000b, pp. 163–4).

The Besley Report avoided recommending the upgrading of the universal service requirements as a solution to these problems. It argued that the universal service regime is a blunt policy tool that can actually limit the operation of a competitive market and be an expensive impost on consumers (DCITA 2000b, p. 172). The report instead supported the moves by the government to open the supply of universal services to competition, and recommended that the upgrading of the universal service requirements be reconsidered at a later stage when some of the new technologies were more firmly established. The Besley recommendations in this regard are consistent with the European Community and Canadian reluctance at this time to upgrade universal service requirements.

5.3 Encouraging Investment in Local Areas

There is a widespread perception in Australia that privatisation and the opening up of the telecommunications market to competition have created inconsistent benefits to consumers. For example, it was argued in submissions made to the Besley inquiry that consumers in the larger cities in Australia enjoyed lower prices and a better range and standard of services, compared to those in regional and remote areas. The policy and regulatory challenge in Australia is to ensure that the benefits flow through to all consumers in a reasonably equitable way. While rejecting the extension of the universal service requirements, the Besley Report recommended that new market models be applied to produce consumer benefits. In research undertaken for the Besley inquiry, the Network Economics Consulting Group (NECG) reported that the private sector, government and community groups were pursuing new delivery models resulting from market opportunities, service gaps and strategic positioning. NECG added that:

> The utility providers, particularly power companies, are recognising the opportunity to provide broadband carriage infrastructure along and through their easements and rights of way. Other companies are providing new data and voice services, in some cases concurrently with the deployment of new backbone infrastructure. A number of rural communities are developing community-based models, some involving partial private sector ownership, while others contract in support services but retain community ownership. (NECG 2000, p. 14)

These developments offer exciting possibilities for both developed and developing nations. They suggest that a dynamic and flexible regulatory approach can encourage and assist service providers, communities and businesses to work at the local level to develop communications infrastructure and services which best respond to local needs. The NECG report found that these developments could be fostered by:

- allowing for different pricing regulation in local areas to ensure there is sufficient profitability for service providers which take on the added financial risks in local areas;
- policies directed at building local knowledge about available telecommunications opportunities, and by encouraging local 'champions' that will offer local leadership and coordination; and
- policies that encourage high standards of governance of the new businesses to manage the risks involved in the new ventures (NECG 2000, p. 28).

An additional policy approach that could be considered is one based on the United States Community Reinvestment Act 1977 (CRA), which applies to the banking system. The problem faced in the United States was that banks were withdrawing their services from low to moderate income population areas, including farming communities. This contributed to the deterioration of these communities, which added to the cost of welfare programmes and increased social problems. The CRA requires the regulators to assess a banking institution's record of meeting the credit needs of its entire community, including low and moderate income neighbourhoods, and publish its CRA performance rating. The rating is taken into account by regulators when assessing certain applications by a bank, including applications for approval of mergers and acquisitions (US Federal Reserve Board 2000, pp. 1–3).

The CRA has led to a substantial increase in lending to lower income populations and neighbourhoods, increased competition, increased demand for credit, and technological advances to assist with making more precise and less costly evaluations of borrower creditworthiness. Many of the lending institutions created special units and programmes to foster CRA related lending. This has led to new and flexible credit products, programmes that educate and counsel prospective borrowers (often linked with government-funded programmes), pricing incentives to mitigate credit risk, enhanced monitoring of repayment patterns, and early intervention programmes to assist borrowers who fail to make timely payments. The Federal Reserve Board of Governors reported, after surveying the operation of the CRA scheme, that community development lending offers a variety of benefits to banks because it promotes community growth and stability, and responds to the credit needs of the local community. Virtually none of the banks reported to the Federal Reserve that they undertake community development lending solely to enhance their CRA rating. Apart from improving the financial standing of the community and the public standing of the banks, the banks reported that on the whole CRA related lending was for the most part profitable or marginally profitable (US Federal Reserve Board 2000, pp. xi, 3–4).

The ratings system could possibly be developed for the telecommunications industry in Australia to provide certain regulatory trade-offs for service providers who developed services and infrastructure to local areas that have service and product levels below that of the rest of the community.

6. CONCLUSION

A superficial examination would suggest that telecommunications reform in Australia since the early 1980s has been successful. The size of the industry has grown dramatically since liberalisation, and consumers have benefited from a greater choice of suppliers and services, and significant price reductions. The lower prices of telecommunications services have flowed through to the whole economy with favourable effects on national efficiency.

Not all of these benefits, however, can be directly attributed to competition. Industry growth, expansion in the range of available services and reductions in prices were all underway prior to reform of the industry, and the more recent results represent to a significant degree the continuation of pre-reform trends in industry performance.

In particular, reductions in the prices of telecommunications services are affected by regulation, technological developments and economies of scale, as well as by competition. It was pointed out in section 4.4 that since 1989, Telstra has reduced its prices by barely more than required by the price caps.

This suggests that it has been regulation and not competition that has been mainly responsible for Telstra's price reductions. As the Chairman of the ACCC has commented, 'the market power of Telstra is such that it could, were its behaviour not constrained by regulation, exert considerable upward pressure on price' (Fels 2001, p. 8). The effect of competition has been to influence the extent to which the regulated overall price reductions have been distributed. They have tended to concentrate on mobile, STD and IDD calls, and to benefit primarily large-volume, high-expenditure customers. While regulation has been largely responsible for the extent of price reductions, competition has caused the reductions to be distributed unevenly across subscriber groups.

This chapter has identified and examined a number of intractable problems in Australian telecommunications. The most important and fundamental of these is the structural imbalance in the industry arising from Telstra's ownership and control of the fixed local access network which gives the company substantial market power in call termination. This market power exists regardless of the number of industry players and Telstra's declining aggregate market share. This is because there is no alternative for call termination other than to use the network operator to which the called party has subscribed for local calls, usually Telstra.

It must thus be concluded that telecommunications reform in Australia has been suboptimal. It has left a structural imbalance that will continue seriously to impede competition and effective regulation of the industry. Regrettably, consideration of measures to overcome this problem of structural imbalance was explicitly excluded by the government from the terms of reference it gave the Productivity Commission for its 2000–01 inquiry into telecommunications competition regulation.

Policy-making in relation to the Australian telecommunications industry can be likened to chasing a kangaroo with a butterfly net. The kangaroo represents the government's policy goal of sustainable competition, and the butterfly net the existing policy measures. The analysis in this chapter, in line with the findings of recent inquiries by both the ACCC and the Productivity Commission, indicates that the policy measures adopted so far have been inadequate to achieve the policy objective for the industry. It is still relatively early days in the reform of Australian telecommunications. Nevertheless, it seems that the kangaroo is unlikely to be secured in the foreseeable future.

NOTES

1. Cable and Wireless Optus was acquired by Singapore Telecommunications in September 2001.
2. Because the Australian telecommunications carriers do not publish data that can be used to calculate precise market shares, it is necessary to rely upon estimates based upon carriers' annual reports and other imprecise secondary data that are produced periodically by various sources including Paul Budde Communication.
3. The data and other information on the various sub-markets set out in sections 2.1 to 2.5 are drawn from Productivity Commission 2001, Chapter 4.
4. Telstra's prices are subject to regulation (see section 4.4).
5. Section 3.4 draws upon and updates the section 'Regulating for Competition' in Brown (1996, pp. 40–9).
6. The approach of the Australian government in introducing telecommunications competition was similar to that adopted in the United Kingdom (see Chapter 2 in this book).

ABBREVIATIONS

ACA	Australian Communications Authority (established 1997)
ACCC	Australian Competition and Consumer Commission (established 1997)
AIEAC	Australian Information Economy Advisory Council
AUSTEL	Australian Telecommunications Authority (1988–97)
CAN	customer access network
CBD	commercial business district
COA-CAM	chart of accounts and cost allocation manual
CPI	consumer price index
CRA	Community Reinvestment Act 1977 (US)
CSPs	carrier service providers
DCITA	Department of Communications, Information Technology and the Arts
DDSO	digital data service obligation
GBEs	government business enterprises
HFC	hybrid fibre coaxial (cable network)
IDD	international direct dialling (calls)
ISPs	internet service providers
NECG	Network Economics Consulting Group
NUSC	net universal service cost
OECD	Organization for Economic Cooperation and Development
OTC	Overseas Telecommunications Commission (1946–92)
PSTN	public switched telephone network
STD	subscriber trunk dialling (domestic long-distance calls)

TCPSS	Telecommunications (Consumer Protection and Services Standards) Act 1999
Telecom	Australian Telecommunications Commission (1975–91)
TIO	Telecommunications Industry Ombudsman
TPA	Trade Practices Act 1974
TSLRIC	total service long-run incremental cost
USL	universal service levy
USO	universal service obligation
USP	universal service provider

REFERENCES

ACCC (1997), *ACCC Directions to the Australian Communications Authority on Number Portability – Commission's Reasoning*, (http://www.accc.gov.au/content/search/index.phtml?filter=1&rowLimit=25&fromItemId=142&searchQueryString=number+portability, accessed 30 September 2001.

ACCC (2000), *ACCC issues final determinations on Telstra access pricing disputes*, Media Release 255/00, 21 September (www.accc.gov.au/media/mr2000/mr-79-00, accessed 25 September 2001).

ACCC (2001), *Review of Price Control Arrangements*, February, (http://www.accc.gov.au/telco/fs-telecom.htm, accessed 1 August 2001).

AIEAC (1999), *National Bandwidth Inquiry*, December (http://www.dcita.gov.au/Article/0,._1-2_1-4_14914.00html, accessed 15 August 2001.

AUSTEL (1994), *Price Control Arrangements for the Australian Telecommunications Industry*, Occasional Paper, Economics 3, June, Melbourne.

Australian National Audit Office (1999), *Sale of One-Third of Telstra*, Audit Report No. 10 (www.anoa.gov.au, accessed 28 September 2000).

Beazley, K. (1991), 'Telecommunications Bill 1991: Second Reading Speech'. Hansard, *House of Representatives*, 7 May, pp. 3093–105.

Brown, A. (1991), 'Golden bird or white elephant? Australia's Aussat satellite system', *Telecommunications Policy*, **15** (3), 248–62.

Brown, A. (1996), 'Reform and regulation of Australian telecommunications', *Australian Economic Papers*, **15** (3), 38–56.

Brown, A. (1997), 'Economic efficiency and public finance aspects of privatisation', *Accountability & Performance*, **3** (2), 65–83.

DCITA (2000a), *Pilot Regions for Universal Service Obligation Contestability: Discussion Paper*, (www.dcita.gov.au, accessed 8 September 2000).

DCITA (2000b), *Connecting Australia: Report of the Telecommunications Service Inquiry* ('Besley Report'), (www.dcita.gov.au, accessed 19 December 2000).

Fels, A. (2001), *Telecommunications: Competition, Effective Regulation and Investment*, ATUG Industry Seminar, 2 October (http://www.accc.gov.au/content/index.phtml/ionternId/179418, accessed 15 September 2000).

Hilmer, F., M. Rayner and G. Taperell (1993), *National Competition Policy*, Report by the Independent Committee of Inquiry ('Hilmer Report'), Australian Government Publishing Service.

Lacy, C. (2000), 'AT&T split shows way of the future for telcos', *Australian Financial Review*, 30 October, p. 11.

Maddock, R. (1992), 'Microeconomic reform of telecommunications: the long march from duopoly to duopoly', in P. Forsyth (ed.) (1992), *Microeconomic Reform in Australia*, St Leonards: Allen & Unwin, pp. 243–59.

NECG (2000), *New Market Models for the Delivery of Telecommunications and Online Services in Regional Australia*, August, (www.telinquiry.gov.au, accessed 26 October 2000).

Paul Budde Communication (1999), *1999/2000 Telecommunications Strategies Australia*, consultancy report, Sydney.

Productivity Commission (1999), *International Benchmarking of Telecommunications Prices and Price Charges*, December, (http://www.pc.gov.au/research/benchmark/telecomsprices/index.html, accessed 22 January 2001).

Productivity Commission (2001), *Telecommunications Competition Regulation*, Draft Report, Canberra, March.

Quiggin, J. (1995), 'Does privatisation pay?', *Australian Economic Review*, **110**, 23–42.

TIO (2000), *Annual Report 1999–2000* (www.tio.com.au, accessed 29 January 2001).

US Federal Reserve Board (2000), 'The Performance and Profitability of CRA-Related Lending', ('CRA Report'), July, (www.federalreserve.gov/BoardDocs/Surveys/CRAloansurvey/, accessed 19 December 2000)

Vickers, J. and G. Yarrow (1988), *Privatization: An Economic Analysis*, London: MIT Press.

Walker, B. and B.C. Walker (2000), *Privatisation: Sell Off or Sell Out? The Australian Experience*, Sydney: ABC Books.

Xavier, P. (1993), 'Price cap regulation for telecommunications: how has it performed in practice?', paper presented to the Conference of Economists, Perth, September.

PART II

ASIA–PACIFIC DEVELOPING NATIONS

4. Indonesia: Telecommunications on a road to reforms

Koesmarihati Sugondo and Risa Bhinekawati

1. INTRODUCTION

The Indonesian telecommunications industry has undergone major changes since the late 1980s. For our purposes, the industry's history can be conveniently separated into three stages, based upon two landmark changes to legislation governing the industry, namely the Telecommunications Act No. 3/1989 and the Telecommunications Act No. 36/1999.

Prior to 1989, Indonesia followed the familiar model of telecommunications services being provided by a government monopoly. The Telecommunications Act No. 5/1964 gave the government the power to both regulate and operate telecommunications. According to the Act, the main objectives of the industry were to serve the national interests, especially in the development of political, economic and military institutions.

Some of the main features of the pre-1989 era were the 1965 separation of the postal and telecommunications services; the establishment in 1967 of Indosat (a subsidiary of ITT) as a provider of international telecommunications services; the creation of Perumtel (a state-owned company) in 1974 as a provider of both domestic and international services; and the sale of Indosat by ITT to the Indonesian government in 1980. Following this sale, there was a separation of the international and national telecommunications services: Indosat became the sole provider of international services, and Perumtel became the monopoly supplier of domestic services.

In this chapter we will examine how the Indonesian telecommunications industry has moved from this position of having two monopoly suppliers to a more competitive environment. Indeed, the regime envisaged in the 1999 legislation will see the government acting as an impartial regulator only, with the commercial operations of the industry being in the hands of the private sector. We will also highlight the main achievements of the reform process to date, as well as the remaining difficulties, and will attempt to draw some lessons from these experiences.

The remainder of the chapter is structured as follows. Section 2 examines the reform process in the 1989–99 period, focusing on the factors driving reform, the major changes involved and their beneficial impact. Section 3 reviews the rationale for further reform and the main features of the 1999 Act, which is expected to take the process much further than before. Section 4 then considers a number of policy issues in this area which Indonesia will need to address in the near future. Section 5 completes the discussion by elaborating some of the lessons to be learned from the reform of the Indonesian telecommunications industry over the last decade.

2. REFORMS IN THE 1989–99 PERIOD

2.1 Drivers of Reform

As telecommunications is an enabling factor for economic growth, the whole Indonesian economy would benefit from a strong and efficient telecommunications industry. Telecommunications also functions as a means of disseminating information, increasing efficiency and productivity of industry, improving the transparency of government administration, enabling distance learning and assisting the provision of health services.

The above arguments apply to Indonesia even more than to most countries, as Indonesia is the world's largest archipelago with more than 17 000 islands and a population of 210 million. The population is unevenly distributed, with 35 per cent of the population living on the island of Java, which contains only 7 per cent of the country's land area, and with the remainder being widely dispersed. Telecommunications can play a very important role in uniting the country. However, financing the construction of adequate infrastructure for such purposes is an enormous task, particularly in the less developed regions of the country.

Due to limited public funding, accelerated telecommunications development can only be achieved by allowing the participation of private and foreign investors, a point well recognised by policy-makers. For example, Jonathan Parapak, former Secretary-General of the Ministry of Tourism, Post and Telecommunications has argued that in order to achieve even relatively modest goals in this area, the country will need to generate much faster growth in the industry. In turn, accelerated growth requires large-scale finance, modern technology and sophisticated human resources. The reality is that the Indonesian government does not have enough of these three factors (Winarno 1999, p. 65).

The increasing demand for telecommunications services, the scarcity of government funds, and the technical difficulties involved in developing, operating and managing modern telecommunications services led the Indonesian government to reform the sector through the Telecommunications Act No. 3/1989. This Act resulted in a shift away from total reliance on government resources toward greater emphasis on private funding.

2.2 Major Changes in the 1989–99 Period

The 1989 Act marked the beginning of the liberalisation of the Indonesian telecommunications industry, as it allowed for the first time the participation of private investors in the provision of telecommunications services. These included both domestic and, to a more limited extent, foreign investors.

Prior to 1989, the development of telecommunication infrastructure was financed primarily from the state budget, often through soft loans from the Overseas Economic Cooperation Fund (OECF) and bilateral foreign assistance under the broad umbrella of the Consultative Group on Indonesia (CGI), previously called the Inter-Governmental Group on Indonesia (IGGI). After 1989, by contrast, the government and its state-owned companies increasingly relied on other means to finance investment in the sector. Indeed, in the early 1990s the government decided that from this period only 30 per cent of telecommunications development would be financed from the state budget, and the rest (70 per cent) was to be financed by the private sector. By 1995, public funding for this purpose had ceased altogether. In what follows, we shall briefly review how the increased participation by the private sector was implemented during the 1990s. The principal methods employed were the partial privatisation of state-owned companies and the formation of strategic alliances with domestic private and foreign companies.

2.2.1 Privatisation of state-owned operators

The partial privatisation of the incumbent operators began in 1991 with the transformation of the domestic carrier, Perumtel, into a state-owned limited-liability corporation with a commercial purpose, named Telkom. A gradual privatisation process then ensued, with shares of both Telkom and Indosat being made available on stock exchanges. In 1994 the Indonesian government obtained US$1,077 million from the initial public offering (IPO) of Indosat, with 25 per cent of the company's shares being sold on the New York Stock Exchange and another 10 per cent on the Jakarta Stock Exchange. This was followed in 1996 by the IPO of Telkom on the New York, Jakarta and London Stock Exchanges. The government, however, remained the majority shareholder of both incumbents. (Even as recently as at the end of 2000, the

Indonesian government still owned 66 per cent of Telkom and 65 per cent of Indosat.)

In privatising Telkom and Indosat, the government was able not only to reduce the need for injections of public funding into major telecommunications investment projects, but also to improve its own financial position. In turn, as a result of the privatisation process, Telkom and Indosat have been repositioned, their competitive edge strengthened, and their ability to form partnerships with international companies improved.

2.2.2 Participation by other operators

Under the 1989 Act, telecommunications services were divided into two types, namely 'basic' services, which included telephone services, telex, telegram and cellular mobile services; and 'non-basic' services, such as paging, trunking and internet services. The Act also delegated the development and operation of telecommunications infrastructure to two state-owned 'Operating Bodies', namely Perumtel (subsequently Telkom) and Indosat. Telkom was the sole operator of *basic domestic* telecommunications services, including fixed local services and fixed long-distance services, and Indosat was the operator of *fixed international* services (its monopoly over this segment of the market was subsequently ended in 1994 following the creation of Satelindo; see below).

Other companies, known as 'Other Bodies', were permitted to operate *non-basic* services independently. But the only way they could operate *basic* services was through cooperation with the Operating Bodies – the incumbents. Moreover, only a limited range of cooperation forms were allowed, the main one being the Revenue Sharing Agreement (more details about this are provided below).

Further liberalisation occurred in the mid-1990s, largely because of the need to keep up with high demand as well as rapid advances in technology. The changes in legislation included Government Regulation No. 8 of 1993 on telecommunication operations, and Ministerial Decree No. 39 of 1993 concerning possible cooperation in the provision of basic services (especially GSM services) in a number of regions. The forms of cooperation permitted within this new legal framework included the Kerja Sama Operasi (KSO) (Joint Operationation Scheme) and joint venture company (JVC). These are discussed in more detail below.

In 1993 the government created Satelindo, which was 60 per cent owned by the Bimagraha Group, 30 per cent by Telkom and 10 per cent by Indosat. The new company took over the management of three domestic satellites that had been under Telkom's control. Satelindo was licensed to provide both international services and national GSM services.

An overview of the structure of the telecommunications industry following these changes is provided in Table 4.1. The Operating Body Telkom was given exclusive rights to operate fixed wireline local telephone services until the end of 2010. Telkom was also granted exclusive rights to fixed national long-distance telephone service until the end of 2005, while rights to operate fixed wireless local telephone services were granted to an 'Other Body', Satelindo. Meanwhile, Indosat and Satelindo formed a duopoly in the market segment for fixed international telephone services. This duopoly is due to expire in 2004.

Table 4.1 Structure of the Indonesian telecommunications industry after 1994

Telecommunications services	Telecommunications operations
Basic telecommunications	
Fixed wireline local services	Exclusive right 1996–2010 (Telkom)
Fixed wireless local services	Limited competition (Satelindo)
Fixed domestic long-distance services	Exclusive right 1996–2005 (Telkom)
Fixed international services	Duopoly 1995–2004 (Indosat, Satelindo)
Mobile cellular services	Competition based on frequency availability
Non-basic telecommunications	Competition exists
Customer premises equipment	Competition exists

Source: Department of Transport and Communications (1999)

While there was little or no competition in the fixed basic services segments of the market, competition in the mobile cellular segment was quite vigorous, due to the presence of a sizeable number of operators. (A more detailed review of this market segment will be presented below.) For non-basic services, which any operator could provide without having to cooperate with the Operating Bodies, licences have been issued to 10 trunking operators, 40 paging operators, 20 value-added services operators, and more than 100 internet service providers. Some sort of competition exists in non-basic telecommunications services. Since these services have been provided by several operators in the marketplace.

As mentioned above, while in theory any company would be allowed to provide basic telecommunications services in cooperation with Telkom or Indosat after 1989, in practice, Telkom was so dominant that there was little room for the others to manoeuvre. It was only after the introduction of new liberalisation measures in 1993, and in particular the subsequent establishment of the Joint Operation Scheme and Joint Venture Company Scheme, that competition became more vigorous and teledensity increased significantly. These schemes facilitated joint public-private participation in the provision of basic services in several regions. They will now be contrasted in turn with the earlier Revenue Sharing Agreements Scheme.

2.2.3 Revenue-sharing agreements

Under a typical Pola Bagi Hasil (PBH) (revenue-sharing agreement (RSA)), Telkom's partners would build infrastructure to be operated by Telkom. In return, Telkom would give a percentage of its revenue to the partners. Examples of these arrangements include those which financed the development of cellular services in the early period.

Cellular mobile services were first introduced into Indonesia in 1986 under a revenue-sharing agreement. A single agent had the monopoly rights to market and sell handsets to customers while Telkom operated the infrastructure. This early revenue-sharing agreement was not a success, as it led to highly-priced terminal handsets, lack of competition, poor service quality and slow demand growth. By contrast, the private participation schemes implemented since 1993 (especially through joint ventures) have had greater success, as we shall see below.

2.2.4 Joint operations schemes

A Joint Operations Scheme allows a licensed consortium to build telecommunications networks and operate services (including those already existing) in a pre-assigned region. The ownership of the assets is to be passed to Telkom eventually (typically after 15 years). In the meantime, Telkom receives a minimum telkom revenue (MTR) which is a fixed income, and a distributed telkom revenue (DTR) which is an agreed percentage (between 30 to 35 per cent) of the remaining revenue after the MTR and Operating Cost have been deducted.

Examples of these schemes include the Telkom joint operations schemes in five of the seven regions of Telkom. Table 4.2 shows the roles played by the strategic partners in the KSO schemes. From the table, it can be seen that the schemes have provided an avenue for foreign telecommunications companies from the UK, France, Japan and Singapore to provide expertise and finance to the Indonesian industry. A cross-section of indigenous private companies has also been represented.

2.2.5 Joint venture company schemes

In a JVC, it is not necessary that one of the Operating Bodies, Telkom or Indosat, be the major shareholder. Under government regulation (PP) No. 20 of 1994, capitalisation of telecommunication enterprises can be conducted by a joint venture company as long as the Indonesian partner contributes a minimum of 5 per cent of the paid capital.

Table 4.2 Ownership of fixed telecommunications operations

Region	Operator	Ownership
Region 1 Sumatra	Pramindo Ikat	Astratel Nusantara (40%) France Telecom (35%) Indosat (13%) Others (12%)
Region 2 Jakarta & environs	Telkom	Govt. of Indonesia (66%) Public, domestic + foreign (34%)
Region 3 West Java	Aria West International	Artimas Kencana (52.5%) Media One (35%) Asian Infrastructure Fund (12.5%)
Region 4 Central Java	Mitra Global Telekomunikasi	Telstra Global (20%) NTT Communications (15%) Indosat (30%) Others (35%)
Region 5 East Java	Telkom	Govt. of Indonesia (66%) Public, domestic + foreign (34%)
Region 6 Kalimantan	Cable & Wireless Mitratel	Intidaya Kharisma (38.7%) Cable & Wireless (25%) Others (36.3%)
Region 7 East Indonesia	Bukaka Singtel	Bukaka Telekomindo (60%) Singapore Telecom (40%)

Source: Compiled from various sources

The cellular mobile operation of GSM 900 is a good example of the evolution of ownership in a JVC scheme. There are three systems of cellular mobile telecommunications in Indonesia. Operation is conducted competitively and the number of operators is limited by frequency availability. The three cellular mobile systems are: the AMPS systems operated by Komselindo, Metrosel and Telesera; the NMT 450 system operated by Mobisel; and the GSM 900 system operated by Satelindo, Excelcomindo and Telkomsel.

The GSM licences were issued to three consortia on a joint venture basis with Telkom in 1994. The licences were for operation in the 900 MHz frequency band with a total bandwidth of 25MHz. All three consortia were allowed to compete both in the operation of network and selling of terminal sets. An 'Open Distribution Channel System' was introduced, and terminal handsets were sold by a chain of dealers and sub-dealers.

The two analogue systems, AMPS and NMT 450, were initially operated under revenue-sharing agreements with Telkom but these were also transformed into joint venture agreements subsequently.

As noted above, Satelindo was originally owned by Telkom (30 per cent), Indosat (10 per cent) and Bimagraha (60 per cent). After foreign strategic partners were allowed to participate in the Indonesian market, shares were bought by DeTeMobil and ownership subsequently changed to approximately Telkom (22.5 per cent), Indosat (7.5 per cent), DeTeMobil (25 per cent), and Bimagraha (40 per cent). Similar changes in ownership occurred with the other two GSM operators, Excelcomindo and Telkomsel. The evolution of ownership in the GSM sector is shown in greater detail in Table 4.3. As of 2001, domestic ownership of Satelindo was approximately 70 per cent.

2.3 Achievements of the 1989–99 Reforms

Indonesia's telecommunications infrastructure and services grew rapidly during the 1990s. As can be seen from Table 4.4, teledensity for the fixed network more than doubled during the five-year period from 1994 to 1998. Other measures of capacity and usage, such as exchange capacity, the number of lines in service, the number of public telephones available and the volume of local and international calls, all grew by similar proportions during this time.

The growth of cellular telephones services was even more spectacular: over the same period, the number of mobile telephones in use increased by fourteen-fold, implying an annual growth rate of around 100 per cent or higher in every year except 1998 (when the country was reeling from the effects of the Asian financial crisis).

The cellular mobile sector clearly illustrated the benefits that competition can bring to the customers, the government, the operators, and the suppliers of telecommunications equipment. The growth of this sector increased significantly after the government granted licences to three digital GSM operators in 1994, eliminated custom duties and opened distribution channels for handsets. There was a threefold jump in the number of cellular telecommunications subscribers in one year, from 78 024 in 1994 to 235 000 in 1995. The liberalisation of restrictions over the provision of mobile cellular services, discussed in relation to JVC schemes above, allowed an increased level of participation by domestic private and foreign investors, the introduction of wider distribution networks for terminal sets, and increased public awareness (Sugondo 1998).

Several cellular mobile companies, especially the GSM operators, invited leading international telecommunications operators to take a minority stake in their companies, generating funding as well as providing access to foreign expertise. Both these factors have accelerated the purchase and rollout of sophisticated IT support, resulting in a rapid growth in the number of subscribers and competition in coverage by the operators.

Table 4.3 Evolution of GSM ownership as a result of the JVC scheme

Name of Operators	Scope of Licence	Original Ownership	Foreign Operators	Ownership as at 2001
SATELINDO	GSM Satellite International Services (1993)	Telkom (30%) Indosat (10%) Bimagraha (60%)	DeTeMobil (1994) US $586 million	Telkom (22.5%) Indosat (7.5%) DeTeMobil (25%) Bimagraha (40%)
EXCELKOMINDO	GSM (1995)	Telekomindo (100%)	NYNEX, Mitsui, AIF (1994) US$297 million	Telekomindo (50%) NYNEX (23.1%) Mitsui (4.2%) Asia Infrastructure Fund Indonesia (12.7%) Santana Telekomindo
(10%)				
TELKOMSEL	GSM (1995)	Telkom (51%) Indosat (49%)	PTT Telecom BV (1996) US$329 million	Telkom (42.72%) Indosat (35%) PTT Telecom BV (17.28%) Setdco Megacell Asia (5%)

Source: Sugondo (1998)

Table 4.4 Development of telecommunications infrastructure and services

Indicator	December 1994	December 1995	December 1996	December 1997	December 1998	December 1999	September 2000
Population (million)	192	195	198	199	204	208	211
Fixed							
Exchange capacity (000)	3 886	4 824	6 343	7 392	8 152	8 359	8 462
Line in service (000)	2 463	3 291	4 186	4 982	5 570	6 080	6 536
Teledensity (%)	1.28	1.69	2.11	2.5	2.7	2.9	3.1
Public phones (000)	81	108	134	166	216	269	328
Production							
Domestic – million pulse	23 419	28 257	35 347	2 124	45 904	47 258	53 054
International – million minutes			577	684	710	645	487
Indosat total	427	492	635	798	837	n.a.	n.a.
Mobile							
Cellular mobile (000)	78	235	562	916	1 085	2 221	3 199
Teledensity (%)	0.04	0.12	0.21	0.46	0.53	1.10	1.50

Source: Telkom and Indosat Annual Reports

Competition induced the operators to introduce new value-added features, such as recorded messages, call line identification, short message systems, fax, data and internet access through handsets. The quality of service also improved.

In short, an important outcome of the reform process has been an increase in competition. With greater competition, the customers enjoy lower prices, better performance, and more choice of services and features, because the operators have to strive to win their loyalty. At the same time, the government also benefits from increased revenues accruing from taxes, frequency fees, and operations fees. (The current operations fee to be paid to the Indonesian government is 1 per cent of gross revenue, and the frequency fee is based on spectrum use as well as traffic.) Similarly, it is expected that rapid telecommunications development will bring opportunities for Indonesia to improve its domestic supply capability. Currently, 85 per cent of telecommunications equipment still needs to be imported because as yet there are not sufficient economies of scale for the country to develop and fully utilise manufacturing capabilities in this field.

3. THE 1999 ACT

3.1 Rationale for Further Reform

The reform process that took place in the 1990s has been strengthened since 1999 by a series of policy initiatives and legislative changes. The impetus for further reform in telecommunications can be found in five areas: the economic crisis of 1998 and the subsequent political changes, continuing regional disparities, remaining inconsistencies and inefficiencies in the system, continuing technological advances; and the requirements of international treaties.

3.1.1 Political support for reform

Indonesia underwent a series of major political, social and economic changes in 1998. The financial crisis in the region, and the consequent crisis in the domestic economy in 1998, resulted in an event of historical importance, in that there was a change in the government that had been in power for more than 32 years. Although the country had to endure many disruptions and difficulties, the socio-political changes which have occurred have also created opportunities for fresh democratic breezes in political life, new ideas for economic development, and the right climate for fundamental reforms.

One of the major decisions made by the new government was to implement regional autonomy and to decentralise public finance. This was enacted in two laws, namely Law No. 22/99 on Regional Autonomy and Law No. 25/99 on the Fiscal Balance between Central and Regional Authorities. Both of these Acts will have an impact on developments in the telecommunications sector, as will the new law on Anti Monopoly and Fair Competition (Law No. 5/1999).

3.1.2 Continuing regional disparities

Despite the gains made in the 1990s, there remained large differences in teledensity between the regions of Indonesia at the end of the decade (see Table 4.5). To a large extent, these reflected the uneven distribution of economic, political and social activities. Regions 2 and 5, which are the centres of political and economic power in Java, enjoyed teledensities well above the rest of Indonesia. In particular, in terms of fixed services, the teledensity of Jakarta (Region 2) in 1999 was more than five times that of Central Java (Region 4). Continuing and accelerated development of the sector was regarded as essential to achieving the goals of decreasing regional disparity and enabling all Indonesians to enjoy the benefits of adequate telecommunications.

Table 4.5 Regional teledensity in fixed services (%)

		1997	1998	1999
Region 1:	Sumatra	1.6	1.8	1.9
Region 2:	Jakarta	7.4	7.9	8.2
Region 3:	West Java	2.0	2.3	2.5
Region 4:	Central Java	1.2	1.4	1.6
Region 5:	East Java	2.4	2.7	3.0
Region 6:	Kalimantan	2.0	2.2	2.4
Region 7:	East Indonesia	1.4	1.6	1.8

Source: DGPT (2000)

3.1.3 Remaining inconsistencies and inefficiencies

Similarly, even though significant advances had been made as a result of the reforms that commenced in 1989, at the end of the 1990s the Indonesian telecommunications industry still faced a range of regulatory problems. In particular, strategic decisions regarding the restructuring of the state-owned incumbents were typically influenced to a large degree by political interests, which were not always consistent with optimal market outcomes. Further, lack of firm policy directions created uncertainty for all key stakeholders, especially the 'Other Body' companies. Investors were hesitant to commit large amounts of funds in the face of uncertainty regarding changes in regulations and tariffs.

3.1.4 Technological change

Technology continued to play an important role in driving telecommunications reform. Pressures from technology developments, such as the convergence of telecommunication and IT services, resulted in a need for new, more streamlined regulatory mechanisms which more adequately take account of recent technological innovations.

3.1.5 International commitments

Another driver of telecommunications reforms was globalisation. Indonesia was required to fulfil its commitments under the 1998 WTO basic telecommunications agreement, which stipulated the safeguards of competition, and the provision of fair interconnection under the transparent supervision of an independent regulator. Similar commitments to certainty, non-discrimination and transparency in telecommunications policies had also been made in regional treaties, such as AFTA and APEC, and with the International Monetary Fund (IMF) through a Letter of Intent which had been signed as part of the rescue package following the Asian financial crisis.

3.2 Major Changes to be Implemented

In 1999 the government issued the Blueprint of Indonesian Telecommunications Reform, prepared by the Department of Communications, and the new Telecommunications Act No. 36/1999, which became effective in September 2000. The intention of the Act is to further liberalise the industry, in order to increase fair and sustainable competition, increase private participation, provide adequate consumer protection, and eliminate regional disparities. These objectives are to be achieved through a separation of the functions of policy-maker and regulator, non-discrimination on the basis of ownership, fair and transparent licensing, cost-based tariffs, fair interconnection, equal access, fair universal service obligations, and technology-neutral standardisation of infrastructure based on international standards.

The major differences between the 1989 legislation and the 1999 legislation are summarised below:

Public–Private mix: Whereas under the 1989 Act, the government reserved a special role for the two state-owned Operating Bodies in the operation of telecommunications, under the new regime services are to be provided by a mix of partly privatised state-owned companies, as well as foreign and domestic private companies. Private companies (formerly 'Other Bodies') will no longer need to cooperate with the 'Operating Bodies' in order to be allowed to provide basic services.

Foreign involvement: Whereas previously foreign ownership of a telecommunications company was restricted to 35 per cent, this ceiling would be gradually raised. For example, in 2000 the foreign ownership limit was increased to 49 per cent with respect to companies providing basic telecommunication networks and services, and 65 per cent for companies providing VSAT services.

Industry structure: Conceptually and operationally, the telecommunications sector is to be divided into network operations and service operations, rather than basic and non-basic services as before. The new classification is less arbitrary and more attuned to the dynamic state of technology in this sector than previously. The proposed structure of the Indonesian telecommunications industry is summarised in Figure 4.1.

SEGMENT						STRUCTURE	OPERATIONAL			
							UNTIL 2004	UNTIL 2005	UNTIL 2010	AFTER 2011
TELECOMMUNICATIONS	PUBLIC NETWORK	SERVICE	MULTIMEDIA	INTERACTIVE MULTIMEDIA		COMPETITION	TELKOM + INDOSAT + OTHER OPERATOR			
				INTERNET		COMPETITION	TELKOM + INDOSAT + OTHER OPERATOR			
			FIXED	FIXED INTERNATIONAL TELECOM SERVICE		Competition (Exclusivity)	INDOSAT SATELINDO	+ TELKOM + OPERATOR LAIN		
				FIXED LONG DISTANCE TELECOM SERVICE		Competition (Exclusivity)	TELKOM		INDOSAT + OTHER OPERATOR	
				Fixed Local Connection Tel. Service	Cable	Competition (Exclusivity)	TELKOM + KSO + BBT			+ Other Operator
					Radio	Competition	Telkom + Other Operator			+ Other Operator
		SERVICE		Value Added Service		COMPETITION	Open			
			MOBILE	GMPCS		COMPETITION	Telkom + Indosat + Other Operator			
				Cellular		COMPETITION	Telkom + Indosat + Other Operator			
				Trunking		COMPETITION	Open			
				Public Pager		COMPETITION				
			Leased Circuit: Satellite Transponder & Resell			COMPETITION	Open			
			Broadcasting			COMPETITION	Open			
						NO COMPETITION	RRI + TVRI			
	SPECIAL TELECOMMU-NICATIONS NETWORK		Close Usage		SRR, PtP	Un-regulated, except frequency usage				
					Other	No Competition	Private			
			Government			No Competition	Government			
			Ministry of Defence & Security			No Competition	Military Purpose			
EQUIPMENT			Non-CPE			Competition	Open			
			CPE							

Source: Department of Transport and Communications (1999)

Figure 4.1 Proposed structure of the Indonesian telecommunications industry after 1994

In the fixed network, Telkom and Indosat are to be repositioned as full service providers, with each being licensed for fixed international, fixed national long distance, and fixed local services. Indosat's and Satelindo's duopoly rights to operate fixed international services will expire after 2004. Similarly, Telkom's exclusive right to operate fixed domestic long-distance services will expire after 2005. Total market liberalisation is scheduled to take place at the end of 2010, when Telekom will lose the monopoly right to provide fixed local services.

In the mobile network, the competitive market structure introduced in the 1990s will be enhanced. As with the fixed network, the restrictions associated with the 'Operating/Other Bodies' distinction have been lifted. In the case of the mobile network, however, there will be no period of exclusivity for the incumbents.

The convergence between telecommunications, broadcasting and information technology is reflected in the new industry structure. The provision of internet and interactive multimedia services will be competitive, and the same will apply to the public network component of broadcasting.

4. REMAINING ISSUES

While the 1999 Act provides a solid framework for reform, in implementing it Indonesia will have to confront a number of challenges, including resistance from the incumbents, political interferences, lack of an established tradition of decisive and independent regulatory bodies, and complex tariff and interconnection issues.

4.1 Cross-Ownership

Cross-ownership within the industry has the potential to inhibit strategic competition. As can be seen from Table 4.3, the two major players, Telkom and Indosat, had joint equity in a number of subsidiary companies. In the case of Satelindo, Telkom had 22.5 per cent and Indosat 7.5 per cent, while with Telkomsel, Telkom's share was 42.7 per cent and Indosat's share was 35 per cent. Such joint equity holdings tend to hamper competition between the two incumbents. More recently, however, the issue appears to have been resolved. Under the terms of a memorandum of understanding signed in February 2001, Telkom will acquire Indosat's 35 per cent interest in Telkomsel; and Indosat will acquire Telkom's 22.5 per cent interest in Satelindo as well as Telkom's 37.66 per cent interest in Lintasarta. (The latter was established in 1998 to provide support to the banking industry, especially in the field of information technology. But before long Lintasarta began to offer data communication services as well). The above developments will give both Telkom and Indosat greater freedom to compete across a broad range of services.

4.2 Resistance by Telkom and Indosat

As may be expected, support for the reform process by the incumbents Telkom and Indosat has generally been lukewarm. In the short run, the reforms simply mean that they will have to give up monopolistic rights and potentially lose market share. Of course, in the long run, the increased level of competition will tend to increase their own productivity and efficiency, and their profitability could conceivably improve (Jonathan Parapak, cited in Winarno, 1999, p. 174). But it is understandable that the incumbents would prefer to face the increased competition later rather than sooner.

Many observers believe that, as far as the ending of the incumbents' monopoly rights is concerned, the reform process can be completed much earlier. However, an acceleration of the process is likely to involve substantial compensation payments to Telkom and Indosat for their short-term losses, at a time when the government is desperately short of funds. Another complicating factor is that the government has an interest in maximising the proceeds from the part-privatisation of Telkom and Indosat, and an acceleration of the de-monopolisation process is not conducive to high capitalisation values for these companies.

4.3 Political Issues

More generally, the management of state-owned companies remains heavily influenced by short-term political decisions, as policy-makers typically see profitable state-owned companies as cash cows. These companies are managed by the Ministry of State-Owned Companies Development, which is subject to heavy lobbying by political parties. Furthermore, there is a lack of awareness among many in the Parliament and the Government of the importance of telecommunications and information technology in national development.

There are also conflicts of interest in implementing regulations such as Presidential Decrees and Ministerial Decrees. As an illustration of this, the supporting regulations had not been made ready by the time the 1999 Act came into effect.

A weak regulatory framework (see below) combined with very strong lobbying from the incumbents make it less likely that difficult decisions will be made, even when they are consistent with the Government's own Blueprint. Lengthy delays in the decision-making process and lack of clear directions make it very difficult for foreign and domestic private companies to develop their future strategies. The continuing uncertainty will hinder the growth of the telecommunications networks and services, to the detriment of the entire country.

4.4 Regulatory Structure

While the 1999 Act and the Blueprint stipulate that there will be a split between the functions of policy-maker and regulator, the notion of an 'Independent Regulatory Body' is a new concept to the Parliament and the Government, with few precedents. It is important to adhere to the concept and ensure that such an agency is established and successfully accommodated within the Indonesian government structure.

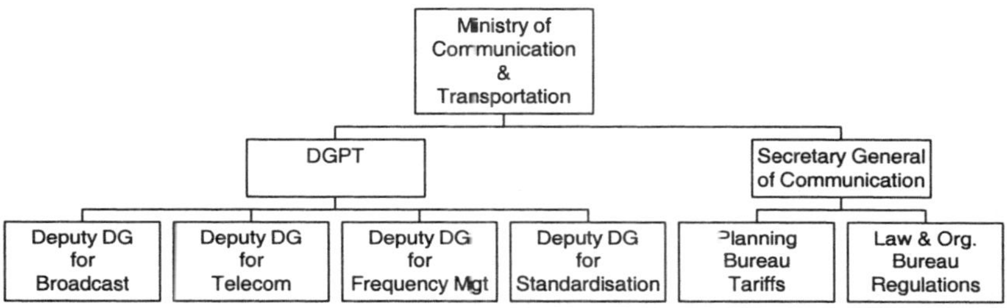

Figure 4.2 Telecommunication regulatory structure in Indonesia

Currently, the Directorate General of Post and Telecommunications (DGPT) of the Ministry of Communications functions as both policy-maker and regulator on behalf of the government. DGPT implements the regulations and supervises companies that have been granted licences to operate telecommunications. After the abolition of the Ministry for Information in 1999, the Ministry of Communications and Transportation established a post of Deputy Director General for Broadcasting within DGPT to accommodate the broadcasting function which had been within the portfolio of the Ministry of Information.

4.5 Tariffs

Future profitability in the sector will depend critically on the government's regulations regarding tariffs and interconnection. From the viewpoint of sustainability, tariffs should be set at a level which is sufficient to ensure the viability of long-term investment. In addition, from the viewpoint of efficient resource allocation, there should be minimal cross-subsidisation across categories of services and users. But the reality is that Indonesian policy-makers have always found it difficult to increase local-service tariffs – which have been among the lowest in the region – due to pressure from the Parliament and the public. This situation can be attributed to both the low purchasing power among most of the consumers and a lack of appreciation of the opportunity and efficiency costs involved.

At present, the government determines maximum tariffs for service components such as new connection fee, monthly access fee, activation charge, etc. It also determines *maximum* tariffs for usage charge, and these typically become the *actual* tariffs – in other words, many of the ceilings are perpetually set so low that they are always binding. The government uses a 'price cap' formula to adjust tariffs periodically, taking into account inflation, productivity and other factors.

Some indicative fixed-line tariffs and charges as at the end of 2000 were as follows: Rp.70 per minute (0.78 US cents) for local calls, Rp.1,589 per minute (20 US cents) for domestic long-distance calls, and Rp.22,300 (US$2.50) for monthly subscription. Most analysts would agree that the local tariff is too low, while the domestic long-distance tariff has to be set artificially high to cross-subsidise the former. Similarly, leased lines are very expensive which is counterproductive to the growth of internet and small/medium enterprises in IT business.

In 1998, although the real value of tariffs decreased by about 70 per cent (due to a 78 per cent surge in inflation following the sharp depreciation of the rupiah) Telkom still managed to report a sizeable profit. This fact has been used by opponents of tariff increases. But it should be noted that a large part of Telkom's revenue was gained in the form of Minimum Telkom Revenue and Distributable Telkom Revenue from its KSO partners under profit-sharing agreements. Telkom's own operations in the fixed-line local call segment were far less healthy in terms of profitability. In essence, Telkom still relies heavily on its exclusivity rights for profitability.

In the mobile cellular market, the government also sets maximum levels for the usage rate per minute, connection fee and monthly subscription fee. These are meant to allow the operators some degree of flexibility to compete in different market segments. However, the ceiling prices, especially the usage rates per minutes, tend to be set so low as to leave the operators little room to manoeuvre. In early 1998, the government granted increases in cellular tariffs of around 28 per cent, but these were not sufficient to allow them to keep pace with general inflation.

Some indicative cellular tariffs set by the government as of 2001 were as follows. The connection fee was Rp.200,000 (US$22) with a monthly charge of Rp.100,000 (US$11). The charge for mobile–mobile local calls was Rp.650 (3.6 US cents) off-peak, with peak-period calls attracting a 25 per cent surcharge. Mobile to PSTN calls cost Rp.390 per minute off-peak and Rp.471.5 per minute during peak periods. These can be compared with the results of a study by the Japanese International Cooperation Agency (JICA) which found that in general cellular operators will need an average revenue per user (ARPU) of around US$225 per year to be able to operate comfortably and reinvest for network and service developments (Sugiyama 1999).

Adjustments in tariffs, such as 'tariff rebalancing' between local and long-distance calls, and an increase in local tariffs are crucial to guarantee the sustainability of Indonesia's telecommunications development. The new Telecommunications Act heralds a more transparent regime of tariffs and interconnection charges, which potentially could provide a level playing field to all players in the industry.

4.6 Interconnections

Both the WTO and APEC agreements, to which Indonesia is a party, stipulate that interconnection charges should be based on cost. In practice, however, this method is difficult to implement as the incumbents, Telkom and Indosat, do not at present calculate disaggregated costs. Further, the interconnection regime is as yet not a level playing field, with the regulator being heavily influenced by the incumbents. To date interconnection charges have tended to be based on extracting maximum profit through monopoly power.

In principle, there are three main ways of calculating cost-based interconnection charges (Ovum 1998, p. 5):

- Bottom-up approach, which involves establishing a network cost model where costs can be calculated based on quotations and references from vendors. This will typically generate very small interconnection estimates, to the disappointment of the incumbents.
- Top-down approach, where calculations are made based on operators' accounts. The results are typically high estimates of interconnection fees, thus favouring the incumbents.
- Benchmarking, which involves international benchmarking of interconnection charges. The difficulty here is to find countries which share similarities with Indonesia in terms of operating conditions, level of economic development, purchasing power, etc.

In practice, a combination of all of these three methods is likely to be needed. At present Indonesia is developing cost-based models which incorporate elements of all three approaches, with a view to devising a new interconnection regime which will be fair for all. In any case, it will be desirable to coordinate interconnection regulations and tariff regulations, and to keep cross-subsidisation to a minimum in order to realise the potential efficiency gains that competition promises to deliver.

5. CONCLUSION

Two lessons are clear from the Indonesian experience with telecommunications reform. First, heavy regulation tends to inhibit rapid growth. Second, there is a need for private (mainly foreign) participation in the industry, both for finance purposes and for the technical expertise that foreign investors bring with them.

It is evident that in the cellular mobile network, where regulation has been less stifling and where foreign and domestic private participation has been more active, growth has been far faster than in the fixed-line market. Indeed, it is expected that the number of cellular subscribers will exceed the corresponding fixed-line number before 2005. In the mobile cellular market, competition has resulted in not only accelerated growth but also more efficient operations and a wider range of offerings to the customer.

While the government has developed sound economic strategies, how vigorous they are in implementing these strategies will determine Indonesia's actual growth path. In particular, despite a commendable vision for the restructuring of the telecommunications sector as detailed in the Blueprint of Telecommunications Reform and the 1999 Telecommunications Act, the authorities continue to face serious challenges in the implementation of the proposed reforms, due to a fundamental conflict of interest between the government's role as the owner of the principal incumbent companies, and its role as the industry's overall regulator. To some extent, the hesitant approach to reform implementation may be attributed to a general lack of appreciation of, and confidence in, the benefits of liberalisation.

The situation can be overcome by thorough discussions, informed with the results of rigorous analysis, and involving not only policy-makers (such as members of the Parliament as well as the Government) but also stakeholders, employees, and members of the public. In addition, the regulating authority should be empowered and transformed into a truly independent body. All these could happen in a relatively short time. What is really needed is the political will to make them happen.

ABBREVIATIONS

AFTA	ASEAN Free Trade Agreement
AMPS	advanced mobile phone systems
APEC	Asia Pacific Economic Cooperation
ARPU	average revenue per user

BAPPEPAM	Badan Pengembangan Penanaman Modal (Agency for Investment Development)
CDMA	code division multiple access
CGI	Consultative Group on Indonesia
DGPT	Directorate General of Post and Telecommunications
DTR	distributed Telkom revenue
GMPCS	global mobile personal communications system
GSM	global system for mobile communications
IGGI	Inter-Governmental Group on Indonesia
IMF	International Monetary Fund
IPO	initial public offering
ITT	international telephone and telegraph
ITU	International Telecommunications Union
JICA	Japan International Cooperation Agency
JVC	joint venture company
KM	contract management
KSO	Kerja Sama Operasi (Joint Operation Scheme (JOS))
MTR	minimum Telkom revenue
NMT	Nordic mobile telecommunications systems
OECF	Overseas Economic Cooperation Fund
PBH	Pola Bagi Hasil (revenue-sharing agreement (RSA))
PP	Peraturan Pemerintah (Government Regulation)
PSTN	public switched telecommunications network
USO	universal service obligations
VSAT	very small aperture terminal
WTO	World Trade Organization

REFERENCES

Department of Transport and Communications (1999), *Government Policy on Indonesian Telecommunications*, also known as *The Blueprint of Indonesian Telecommunications Reform*, Jakarka.

DGPT (Director General of Post and Telecommunications), (2000), *Strategic Development Plan in Post and Telecommunications*, Jakarta.

European Commission (1998), *Case Study of the Changing International Telecommunications Environment – Indonesia*, Jakarta.

Ovum, (1998), *Implementing Cost Based Interconnection*, Consultative Report, Jakarta.

Sugiyama, H. (1999), 'Long term development plan – shift from fixed telephone', paper presented to the Minister of Transport and Communications, Jakarta.

Sugondo, K. (1998), 'Financing cellular ventures in Indonesia', paper presented to the Cellular and Wireless Conference, London, July.

Sugondo, K. (2000), 'Direct (equity) foreign investment in telecommunications – Indonesian case', paper presented to the Conference on Telecommunications Reform in the Asia-Pacific, Jakarta, November.

Winarno, B. (1999), 'Creating value in a state owned company – a case study of PT Indosat', pp. 165–74.

5. Malaysia: Telecommunications reform and beyond

Cassey Lee

The telecommunications sector in Malaysia has undergone significant physical and structural transformation since the early 1990s. Between 1985 and 2000, the country's telephone penetration rate rose by 540 per cent. Equally important, the liberalisation and privatisation of the sector in the 1980s ushered in an era of regulatory reforms and competition in the sector. The market structure as well as the regulatory framework and the institutions for the telecommunications sector continue to evolve. Other nations that have privatised their telecommunications have now realised, privatisation *per se* is easy. The real challenge, however, lies in what to do after that – putting in place an adequate regulatory framework and institutions that will ensure industry growth as well as protect consumer interest. The on-going microeconomic and regulatory reforms in this sector seek to fine-tune the regulatory mechanisms over time.

This chapter reviews the recent history and development of the telecommunications sector in Malaysia. Section 2 provides a brief historical account of the sector and the current structure of the sector. This is followed by a discussion on regulatory reforms in section 3. Section 4 examines the impact of reforms in the telecommunications sector. Section 5 concludes by discussing the future policy agenda for the sector.

1. TELECOMMUNICATIONS INDUSTRY STRUCTURE

1.1 Physical Expansion

The infrastructure sector plays a key role in Malaysia's economic growth and development. This sector's share of development expenditure in the various five-year plans implemented since 1966 has been between 18 and 34 per cent (Lee 2000). The emphasis on infrastructure development in Malaysia has been very strong. The high rate of expenditure appears to be targeted mainly at the transport and the power sectors. As a result, telecommunications indicators – such as the telephone penetration ratio – remained relatively low particularly before 1980 (Table 5.1). In 1970, the fixed line penetration ratio was merely 1.3 per 100 population. This figure increased to 2.9 per 100 population in 1980 and 9.2 per 100 population in 1990. Significant gains in the fixed line penetration ratio were made after telecommunications reforms were undertaken at the beginning of 1987. By 1995 and 2000, the fixed line penetration ratio had risen to 16.6 per 100 population and 21.0 per 100 population respectively (note: the direct exchange line penetration rate is around 20 per 100 population). Concomitantly, cellular phone subscriptions have grown very rapidly since the early 1990s. The total number of cellular phone subscribers reached close to 78 000. By 2000, this had increased to 5.1 million.

Despite the impressive achievement in telephone penetration (for both fixed line and cellular telephones), access to telecommunications services remains uneven. The penetration rates in some of the poorer states such as Kelantan, Sabah and Sarawak are significantly below the national average (Table 5.2).

1.2 Liberalisation and Privatisation

Reforms in the infrastructure sector in Malaysia began in the early 1980s. The experience of twin deficits and mounting external debt during this period convinced the Malaysian government to embark on a strategy of promoting private-sector-led growth and development. The government considered privatisation to be a way to relieve their administrative and financial burdens and at the same time improve the efficiency in service delivery. Privatisation was also considered to be an important avenue to address the issue of economic imbalance between the different races in the country. More specifically, the government sought to increase the participation of the indigenous community (or 'Bumiputeras') in mainstream economic activities. Thus, reforms that were carried out in Malaysia had both efficiency and equity considerations. It is difficult to ascertain the extent of conflict between these two objectives. Some critics tend to think that distributive considerations were given more emphasis compared to efficiency objectives.[1]

In the telecommunications sector, reforms began in 1983 when the government allowed the private sector to supplement Jabatan Telekom Malaysia (JTM) in the supply of terminal equipment such as telephones and teleprinters. This initial effort to liberalise the telecommunications sector was driven by a shortage of terminal equipment as well as the private sector's ability to convince the government they could remedy the problem.

Table 5.1 Malaysia – telecommunications indicators, 1970–2000

	1970	1980	1985	1990	1995	2000
Population size ('000)	8 118	13 879	15 883	18 102	20 689	23 250
Fixed line subscribers	107 000	395 640	948 598	1 585 744	3 332 447	4 430 799
Cellular phone subscribers	0	0	4 630	84 557	872 790	5 100 000
Total telephone subscribers	107 000	395 640	953 228	1 670 301	4 205 237	8 930 799
Telephone penetration ratio (per 100 population)	1.3	2.9	6.0	9.2	20.3	38.4
Fixed line penetration ratio (per 100 population)	1.3	2.9	6.0	8.7	16.6	21.0
Cellular penetration ratio (per 100 population)	0	0	0.03	0.5	4.2	19.3
Rural penetration ratio (per 100 population)	n.a.	n.a.	n.a.	2.0	5.5	11.7
Urban penetration ratio (per 100 population)	n.a.	n.a.	n.a.	n.a.	24.8	28.6

Sources: Sixth Malaysia Plan; Mid-term Review of the Sixth Malaysia Plan; Seventh Malaysia Plan; Mid-term Review of the Seventh Malaysia Plan; Eighth Malaysia Plan

Table 5.2 Malaysia – telephone penetration ratio, by state, 1998

State	Population (1000)	GDP per capita (RM)	Telephone Subscribers			Telephone penetration ratio
			residential	business	total	
Johor	2 611.6	12 705	447 525	126 054	573 579	21.9
Kedah	1 554.8	8 896	249 975	49 864	299 839	16.9
Kelantan	1 484.1	4 293	112 639	21 320	133 959	9.0
Melaka	587.6	12 785	108 108	32 530	140 638	23.9
N. Sembilan	823.4	12 245	150 270	33 189	183 459	22.3
Pahang	1 264.8	10 784	145 179	34 295	179 474	14.2
Perak	2 106.4	11 682	371 934	79 991	451 925	21.5
Perlis	221.8	8 724	–	–	–	–
P. Pinang	1 234.4	17 509	254 660	91 976	346 636	28.1
Sabah	2 740.0	7 028	154 638	70 723	225 361	8.2
Sarawak	1 990.2	11 108	188 705	78 892	267 597	13.4
Selangor	3 092.8	17 708	754 466	315 880	1 070 346	34.6
Terengganu	1 004.1	20 623	100 031	22 143	122 174	12.1
Kuala Lumpur	1 390.8	26 442	146 430	182 288	328 718	23.6
Malaysia	22 106.6	12 884	3 184 560	1 139 145	4 323 705	19.6

Sources: Department of Statistics (1999a and 1999b); Mid-term Review of Seventh Malaysia Plan

Several locally-owned as well as foreign companies took advantage of the new market environment. This set the stage for further liberalisation in other telecommunications markets such as value-added network services (VANs) (1984), radio paging (1985), and mobile cellular services (1988).

Even as the telecommunications sector was being liberalised in the early 1980s, plans were already at hand to privatise JTM (Kennedy 1995). By 1985, a series of legislative changes were carried out to make this possible. Two years later, on 1 January 1987, Telekom Malaysia (TM) officially took over the operational responsibilities of JTM. Subsequently, the government sold 25 per cent of TM's equity to the public via a public listing exercise in 1990. (With the public listing, TM became Telekom Malaysia Berhad or TMB). To date, the government continues to own about 70 per cent of TMB's equity through various agencies (Telekom Malaysia Berhad 2000).[2] In addition, the government owns a single unit 'golden share' which entitles it a veto on major decisions of the company. This arrangement is common amongst major privatised utilities, for example, Malaysian Airlines (MAS) and Tenaga Nasional Berhad (or TNB, the incumbent power utility).

After the privatisation of TMB, the government continued to liberalise the telecommunications market despite having substantial shareholding interests in the incumbent firm. Kennedy (1990, 1995) and Gomez and Jomo (1999) suggest that political patronage was responsible for the liberalisation of the sector before and after the privatisation of JTM. Liberalisation in the post-privatisation era mainly took the form of entry into two major markets, fixed line and cellular services. Between 1993 and 1995, five additional licences to operate in the fixed line market were approved (Table 5.3). Despite the increase in the number of participants in the fixed line market, the incumbent's (TMB) dominance remained unchallenged – its market share in 2000 was estimated to be as high as 96.7 per cent (TIME dotCom 2001, p. 26). The prohibitive cost of building a fixed line network from scratch discouraged other licence holders from competing with TMB in this market. Furthermore, the advent of cellular services offered greater growth opportunities at lower investment costs compared to fixed-line services.

The lower start-up cost in terms of investment in the cellular phone services market compared to the fixed-line market heralded the licensing of new competitors in the market. The first cellular phone services (the NMT450) in Malaysia was introduced in 1984. Subsequently, a second licence was issued in 1988 to an STM subsidiary (STM Cellular Communications Sdn Bhd) to provide cellular phone services based on a more advanced technology (ETACS ART 900). In 1989, STM sold 51 per cent of its share in the subsidiary company to Alpine Resources Sdn Bhd. In the following year, STM sold its remaining 49 per cent share in the company. Following this event, the company was renamed Celcom Sdn Bhd.

Table 5.3 Major telecommunications markets in Malaysia, 2000

Market	Operator	Year licence issued	Year operation started	Estimated number of subscribers in 2000
Fixed line	Telekom Malaysia Berhad	1987		3.5 million
	Maxis Communications Sdn Bhd	1993		30 000
	TIME dotCom Berhad	1994		5 000
	PrismaNet (M) Sdn Bhd	1994		4 000
	DiGi Telecommunications Sdn Bhd	1995		–
	Celcom (M) Sdn Bhd	1994		–
Cellular service	Telekom Malaysia Berhad (Atur, Mobikom & TM Touch)	1985	1985	671 000
	Celcom (M) Sdn Bhd (Art900 & Celcom GSM)	1989	1990	1 340 200
	Maxis Communications Sdn Bhd (Maxis)	1993	1995	1 207 000
	DiGi Telecommunications Sdn Bhd (DiGi)	1994	1995	805 000
	TIME dotCom Berhad (TIMECel, formerly ADAM)	1993	1995	447 000
Internet service provider	Telekom Malaysia Berhad (TMNet)	1996	1996	350 000
	MIMOS Berhad (Jaring)	1987	1992	800 000
	Maxis Communications Sdn Bhd (Maxis Net)	1998	2000	300 000
	TIME dotCom Berhad (TimeNet)	1998	2000	137 000
	Celcom (M) Sdn Bhd (Celcom.net)	1998	2000	–
	PrismaNet (M) Sdn Bhd	1998	–	–
	DiGi Telecommunications Sdn Bhd (DiGi)	1998	–	–
	NTT MSC Sdn Bhd (Arcnet)	2000	2000	2 000

Sources: Interviews; Media reports

Although (after the first year of operation) the incumbent firm had twice the number of subscribers (57 000 in 1990) than the other firm (Alpine Resources, which had 23 000 subscribers in 1990), its subscriber base was quickly overtaken, within two years. By 1992, Celcom's subscriber base was 123 330 against TMB's 83 118. TMB's market share continued to decline after an additional six licences were issued in 1993 and 1994. TMB's market share would have been even worse had it not acquired two operators in the market (MCRB in 1996 and Mobikom in 1998).

Aside from market entry via the issuance of licences, technological change has been an important contributing factor to the intensification of competition in the cellular phone service market. Prior to 1995, cellular phone services offered by TMB (ATUR450), Celcom (ART 900) and Mobikom were first-generation platforms featuring analogue voice services. In 1995, the introduction of second-generation platforms featuring digital voice services essentially created a more level playing field in the market. TMB's GSM-based service (TM Touch) had 2513 subscribers in 1995, a figure substantially less than other operators offering digital cellular phone services. For example, Celcom's GSM services had 25 820 subscribers, Maxis (formerly known as Binariang) had 25 820 and DiGi (formerly Muitara Telecommunications), 23 878 in 1995. By the year 2000, both Maxis and DiGi had increased their market share in the cellular phone services market substantially to 26.5 per cent and 17.3 per cent respectively. The market shares of both incumbents (Celcom and TMB) were eroded significantly (see Table 5.4).

The significant market share gains made by upstarts such as DiGi were partly due to the sale of pre-paid cellular telephony services. DiGi was the first firm to launch pre-paid services in January 1998. Not long after this, in the same month, Celcom introduced their own pre-paid services. Other companies followed suit – TMB's TMTouch in December 1998 and Maxis in October 1999. Despite these developments, DiGi appears to have benefited from the first move advantage – its market share of the pre-paid market was estimated to be more than 40 per cent in year 2000.

Another important telecommunications market that has been liberalised is that of the internet service provider (ISP). The incumbent in this market is MIMOS Berhad, originally a government-owned research institute that was established to promote research and development in the electronics and communications industries.[3] MIMOS began providing an internet service in 1987. Nine years later, a second ISP licence was issued to TMB. In 1998, five additional licences were issued to the other major firms in the sector (TIME, Maxis, Mutiara, Celcom and Prismanet). Of these, three companies (TIME, Maxis and Celcom) began offering ISP services in 2000. The dominant ISP firms are MIMOS and TMB with a combined market share of more than 70 per cent in 2000.

Table 5.4 Malaysia – market shares in the cellular phone market, 1990–2000

Company	Number of subscribers (000)		
	1990	1995	2000
Telekom Malaysia Berhad	57	99	948
(%)	71.3	11.5	18.5
Mobikom		150	(part of TMB)
(%)		17.5	
Celcom (M) Sdn Bhd	23	561	1 530
(%)	28.7	65.2	29.9
Maxis Communications Sdn Bhd	–	13	1 359
(%)		1.5	26.5
DiGi Telecommunications Sdn Bhd		24	884
(%)		2.8	17.3
TIME dotCom Berhad		13	401
(%)		1.5	7.8
Total	78	860	5 122

Source: Communications and Multimedia Commission (CMC)

2. REGULATORY REFORMS

The two major events that have significantly transformed the regulatory regime for the telecommunications sector in Malaysia have been privatisation in 1987 and the enactment of the Communications and Multimedia Act in 1998.

2.1 Privatisation and the Change in the Regulatory Regime

Prior to 1987, telecommunications services were provided by Jabatan Telekom Malaysia – a government department under the Ministry of Energy, Telecommunications and Posts (METP). As such, the sector was self-regulated with JTM receiving instructions directly from the ministry. Privatisation brought about a fundamental change in the regulatory regime in telecommunications.

First, the Telecommunications Act of 1950 was amended to make JTM the regulatory authority. Decisions pertaining to licences (an important market entry condition) continued to be made by METP (as provided for in the Telecommunications Act). Second, the passage of the Telecommunications Service Act of 1985 paved the way for Syarikat Telekom Malaysia (STM, the precursor of TMB) to take over the provision of telecommunications services from JTM, which it did in 1987. With these developments, the provision of telecommunications services and regulatory functions were, for the first time, assumed by different organisations.

In 1994, the Ministry of Energy, Telecommunications and Posts issued a 28-page policy paper for the telecommunications sector, the *National Telecommunications Policy* (NTP). Essentially, NTP contains a set of policy recommendations – both at the macro (general) and micro (specific) levels – aimed at developing the sector between 1994 and 2020. The termination year for the policy (2020) coincides with the target date for the achievement of developed country status by Malaysia – a policy objective dubbed 'Vision 2020' that was announced by the Prime Minister Mahathir Mohamad in February 1991.

In many respects, the NTP foreshadowed some of the important changes that were going to take place in the regulatory policies of the sector. It is explicitly stated in the NTP that competition is an important dimension of regulatory policy:

> The NTP's main approach is to encourage a healthy and orderly competition. ... The main objective of the telecommunications sector is to encourage competition in the telecommunications sector in order to achieve efficiency and to provide excellent and quality service. (*National Telecommunications Policy* 1994, p. 9)

Despite this emphasis on the importance of competition, the government did not subscribe to a totally laissez-faire approach:

> Even though competition is encouraged, the Government is empowered to determine the number of competitors that are economically viable for certain telecommunication systems/services. (*National Telecommunications Policy* 1994, p. 10)

Another important aspect of the NTP is how the industry structure is perceived within the regulatory framework. The industry structure is described in the NTP as comprising two major components:

- *Network infrastructure*, made up of:
  (a) basic network infrastructure and
  (b) value added network infrastructure; and
- *Telecommunications services*, made up of:
  (a) basic telecommunications services and
  (b) value-added telecommunications services.

This view was consistent with then prevailing licensing practices where licences were service-specific (telecommunications, cellular, internet) and technology-specific (for example, AMPS, GSM, PDC in the cellular market). However, with the passage of the Communications and Multimedia Act 1998 (CMA) four years later, the licensing structure was to change. Aside from this, CMA brought about a more formal approach to the treatment of competition in the new regulatory structure.

2.2 The Communications and Multimedia Act 1998 and the Communications and Multimedia Commission

In November 1998 the Malaysian Government restructured the Ministry of Energy, Telecommunications, and Post into the Ministry of Energy, Communications and Multimedia (MECM). A major reason for this restructuring exercise was to bring the regulatory structure into line with technological developments, in particular the convergence in communications and multimedia industries. (The Ministry defines convergence as 'the progressive integration of the value chains of traditional communications and content industries within a single value chain based on the use of distributed digital technology' http://www.ktkm.gov.my).

Concurrent with this restructuring exercise, a new regulatory authority for the sector, the Communications and Multimedia Commission was formed. The commission is comprised of five members. Even though only a single member of the commission is a government representative, all five are appointed by the Minister of Energy, Communications and Multimedia. In April 1999, the CMC assumed the regulatory mandate for the sector with the passage of both the *Communications and Multimedia Act 1998* (CMA 1998) and the *Malaysian Communications and Multimedia Commission Act 1998* (CMCA 1998). The former spells out the regulatory institutions and structure for the industry while the latter details the function, powers and operational aspects of the CMC.

The CMA 1998 is a comprehensive piece of regulatory legislation. Its enactment streamlines the regulatory structure by consolidating the many legal statutes (Telecommunications Act, Broadcasting Act) and regulatory authorities (METP, Ministry of Information, JTM) for broadcasting and telecommunications by putting them under a single umbrella regulatory framework.

Under the present regulatory framework, the Ministry of Energy, Communications and Multimedia is the most influential institution (see Figure 5.1). The Minister responsible makes all the key decisions pertaining to regulatory policies (such as licences, directions, determinations and declarations). The CMC provides recommendations to the Minister and is responsible for the enforcement of regulatory policies and legislation (the CMA 1998). In the case of the issuance of licences, the Commission administers the application and renewal process, makes recommendations but the final decision remains with the Minister.

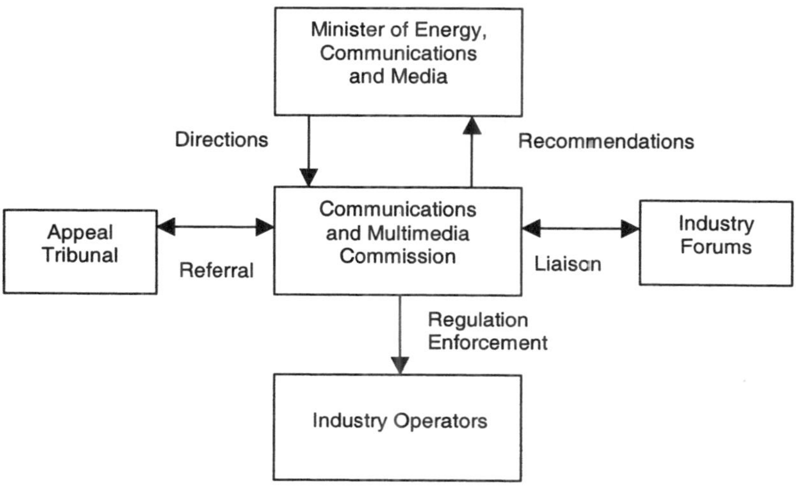

Sources: Adapted from CMC (2001a); CMA (1998)

Figure 5.1 The regulatory framework for the telecommunications industry in Malaysia, 2000

There is a greater role for the public to participate in the present regulatory framework. The CMA 1998 provides for the conduct of public inquiries by the CMC on regulatory matters. The CMC has used this avenue to solicit opinions from operators during the process of drafting regulatory policies (for example, Access List Determination). Typically, discussion papers are published on the internet, and the industry is invited to submit their views within a given period (at least 45 days). The CMA 1998 also allows for the set-up of an Appeal Tribunal to review CMC decisions and direction when the need arises. Another avenue for public participation is the establishment of industry forums that act as consultative bodies to the CMC in important issues such as access code, consumer code, content code and technical code. Thus far, two industry forums have been established namely the Consumer Forum and the Content Forum.

Even though the CMC takes directives from and makes recommendations to the Minister of Energy, Communications and Multimedia, it is an organisation outside the Ministry. Unlike its predecessor, JTM, this set-up provides the CMC with greater flexibility with regards to regulatory initiatives. This is crucial as it embarks on new functions such as monitoring competition. Besides, as a statutory body, the CMC also has greater flexibility in staff recruitment. In particular, since its establishment, it has adopted a 'greenfield' approach to staff recruitment – that is, hiring people from the private sector, the industry and other regulatory bodies. This has created opportunities to build its regulatory capacity independent of government influence.

2.3 Regulatory Principles

The CMA 1998 has substantially clarified and deepened the regulatory framework for the telecommunications sector. Under the Act, regulatory activities are classified into four key areas: economic regulation, technical regulation, consumer protection, and social regulation (see Table 5.5).

Table 5.5 Key regulatory principles under the Communications and Multimedia Act, 1998

Key functions	Objectives
Economic regulation	Industry players must have incentives to invest, innovate and interconnect for the benefit of end-users. There should also be incentives to encourage the export of services.
Technical regulation	Networks and services must be inter-operable, safe, secure, reliable and guarantee the integrity of services delivered.
Consumer protection	Consumers have rights to high quality service that are reliable, easily accessible, and affordable from the service-provider of their choice.
Social regulation	Content developers must have incentives to invest and innovate in applications and services that promote Malaysian culture, identity and values.

Source: CMC (2001a)

2.3.1 Economic regulation

The major objective of economic regulation is to ensure that the telecommunications industry is efficient. This is to be achieved through the implementation and enforcement of three types of regulatory policies – licensing, competition policy and service access.

The licensing structure underwent significant changes under the CMA 1998. Previously, licences were issued for specific technologies and services such as telecommunication, broadcasting (TV or radio), cellular, ISP and VAN licences. Under the CMA 1998, licences were issued for four major categories of activities, namely content application services, application services, network services, and network facilities. This new licensing structure takes cognisance of the potential for unbundling communications services vertically, allowing competition at different levels along the upstream–downstream continuum of the industry (akin to enhancing horizontal competition). The activity-based licensing is also expected to facilitate the introduction of convergent services.

Prior to the CMA 1998, previous telecommunications legislation (the Telecommunications Act 1950) did not have any provisions on anti-competitive conduct. Recognising the importance of market competition in the achievement of efficiency and in the absence of a national competition law, the CMA 1998 took the bold step of addressing this issue by incorporating in the Act prohibitions of anti-competitive practices. Since its enactment, the CMC has published a series of guidelines on relevant concepts such as 'substantial lessening of competition' (CMC 2000a) and 'dominant position' (CMC 2000b). It has also published guidelines on the procedures and processes for assessing and remedying anti-competitive conduct (CMC 2000c).

Access to essential facilities such as the fixed network is an important pre-condition in ensuring level-playing field competition in the telecommunications market. To this end, the CMA 1998 provides for the determination of an access list. It was completed and came into effect in April 2001. Having settled the issue of an access list, the CMC is currently looking into the issue of access pricing.

2.3.2 Consumer protection

Consumer protection is undertaken to ensure that consumers' needs in terms of access, affordability, and service quality are met. To meet these objectives, the CMA 1998 provides for the implementation of regulatory policies and, if necessary, institutions to deal with key issues such as service quality, required applications services, consumer disputes, rate regulation and universal service provision. The two areas that have received the most attention from the public are rate regulation and universal service provision.

The CMA 1998 has provisions for market-based rate setting (that is, cost-oriented and no cross-subsidies) but ironically it also provides for the Minister to intervene on this matter. The latter tends to dominate the former – thus tariff re-balancing requires Ministerial approval. Equal Access by way of call-by-call – which enables subscribers to choose their long distance (STD) or international (IDD) carrier via dialling of a 3-digit selection code – was implemented in January 1999.[4] An important aspect of the implementation of Equal Access is the ruling that operators are not allowed to offer rate discounts that undercut TMB's published rates by more than 20 per cent. And, an even more ambitious version of Equal Access, by way of pre-selection – which allows subscribers to permanently pre-select the long-distance or international carrier without dialling a selection code – has been deferred.[5]

The CMA 1998 provides for the establishment of a Universal Service Fund (USF) to improve network facilities and services in under-served areas and for under-served groups within the community. Prior to 1999, TMB was the only network operator required to fulfil universal services obligations (USO). In the mean time (for 1999 and 2000), other network operators were required to contribute to the provision of universal services in proportion to their revenues and weighted by services (CMC 2001a). The USO regime in Malaysia is currently undergoing transition to a full USF-based system.

Prior to the CMA 1998, provisions on service quality and consumers' needs were subsumed under very brief objective statements (for example, 'the Director General shall have regard to – (a) efficiency and economy; (b) satisfying all reasonable demands for telecommunication services ...' (section 3B, (2) of the Telecommunications Act 1950)). The CMA 1998 makes explicit provisions for the formation of a Consumer Forum (established in February 2001) to deal with matters pertaining to consumer protection.

2.3.3 Technical regulation

Technical regulation is important to ensure technical inter-operability of networks, efficient allocation of resources and the safety, security and integrity of network services and applications services (Syed 1999). The key areas under technical regulation are spectrum assignment, numbering and electronic addressing, and technical standards. The CMC can issue spectrum assignments to operators but only after the Minister has determined the frequency bands for it. Once again, this put the Minister in a powerful position in spectrum assignment.

2.3.4 Social regulation

Social regulation is meant to ensure that content applications and services reflect national cultural aspirations. This area of regulation is targeted at offensive and ant-social contents. There is no mention of Ministerial intervention in this matter but the Minister's control over licences certainly provides a strong case for some form of moral suasion. The CMA 1998 also has provisions for the promotion of public education. As provided by the

CMA 1998, an advisory and consultative body in the form of the Content Forum was formed in February 2001.

3. THE IMPACT OF REFORMS

The impact of reforms has to be evaluated against the objectives of implementing them, namely efficiency and equity. Based on several indicators (such as return on assets, revenue per subscriber, production per employee, direct exchange lines per employee and response rate to customer complaints), TMB's financial performance seems to have improved after it was privatised in 1987 (see Table 5.6).

Table 5.6 Performance of Telekom Malaysia Berhad before and after privatisation

Indicator	Before privatisation (circa 1986)	After privatisation (1997)
Return on assets (%)	4.0	7.6
Average revenue per user (RM)	1 227	1 609
Production per employee (RM)	34 372	219 641
Number of direct exchange lines per employee	36	154
Percentage of responses to customer complaints within 24 hours	80	91.5
Operating revenue (RM million)	1 644.2	7 165.7
Profit before tax (RM million)	4.9	2 376.4
Debt equity ratio	2.3	0.7

Sources: Mid-term Review of Seventh Malaysia Plan; Poon (2000)

But these are achievements made over a period of ten years. Revenue per subscriber, for example, seems to have increased by 30 per cent during this period. This is not substantial in relative terms, if one considers the fact that between 1987 and 1997, TMB's operating revenue grew at an average annual rate of close to 14 per cent. Still, substantial improvements were made in productivity. For example, production per employee rose by 6.4 times its value in 1987.

Since its privatisation in 1987, TMB has faced competition from other operators in two areas – cellular mobile phone services and the long-distance and international call market. Competition in the cellular services market increased with the entry of three major operators (Maxis, DiGi and TIME) in

1995. The implementation of Equal Access in January 1999 also paved the way for competition in the long-distance and international call market. Have consumers benefited from these developments? In the cellular mobile services market, there is some evidence that rates have declined. The average revenue per unit (ARPUs) is reported to have declined from RM150 to RM120 for post-paid services and from RM90 to RM60 for prepaid services.[6] Competition in the long-distance and international call market is also driving rates down but this has been limited by the maximum discount ruling (not more than 20 per cent off TMB's rates). TMB, for example, offered discounts between 15–20 per cent on its national and international calls in June 2000. A month later, Celcom too began offering discounts on its domestic and international calls (*Malaysian Business*, 1 March 2000).

Telecommunications operators will continue to invest in the sector only if they are assured that they are able to recoup costs as well as earn a reasonable rate of return on their investments. Does the telecommunications sector in Malaysia fit this description? In the presence of high start-up costs, ever-changing technology and over-capacity, most of the operators are heavily indebted. Total corporate debt in the sector is estimated to be in excess of RM12 billion (*Business Times*, 9 May 2001). In the past, these companies have sought foreign partners to meet the substantial investment cost as well as the expertise requirement. Taking cognisance of this, the government raised the limit on foreign shareholding in the sector to 61 per cent. At present three of the major operators have foreign partners: Maxis (33 per cent share owned by British Telecom), Celcom (21 per cent by Deutsche Telekom AG) and DiGi (30 per cent by Telenor International AS).[7] This pragmatic policy has meant that the equity aspect of the telecommunications reforms has taken a back seat.

4. FUTURE POLICY AGENDA

Market forces were brought to bear on the sector with liberalisation and privatisation policies in the 1980s. A decade later, in 1998, the *Communications and Multimedia Act 1998* (CMA 1998) marked a new phase in telecommunications reform in Malaysia – that of strengthening the regulatory structure and institutions in the sector. A plethora of regulatory issues continue to be addressed – competition, access, universal services, among others. But numerous policy issues remain unresolved in the telecommunications sector in Malaysia.

A fundamental principle underlying the CMA 1998 is transparency. Public participation through public inquiries and industry forums is a good start. However, it is still unclear whether a more extensive notion of transparency is desirable and possible since key regulatory policies and decisions are ultimately made by the Minister of Energy, Communications and Multimedia.

The CMA 1998 provides for market-based rate-setting but this has not been implemented. As long as rates are set by political masters without them having to justify their decisions it will be difficult to deal with rate-setting issues such as rate-rebalancing, equal access and access pricing. The benefits of competition in the form of lower prices will not be forthcoming when constraints on price flexibility remains. The choice of a rate-setting system will also have implications for productivity improvements (for example, via incentive regulation) and investment incentive in the industry.

Competition policies and laws are virtually non-existent in Malaysia. The incorporation of competition issues in the CMA 1998 is an attempt to remedy this. The lack of tradition and experience in this area means that it will be difficult for the regulator (CMC) to take on (as it has) the responsibility for addressing anti-competition conduct in the sector. The government has to consider the option of implementing a national competition law to deal with anti-competitive conduct rather than having a individual regulatory commission for each sector (telecommunications, power, ports and so on).

Market structure and the role of foreign investment in the sector will be important issues in the coming years. With the third-generation cellular technology (G3) on the horizon, the industry will require additional infusion of funds. With most of the existing operators being heavily indebted and the industry facing diminished foreign participation, technological diffusion may be delayed. In this regard, the issue of market structure will emerge again, that is, determining what is the optimal number of operators to ensure investment cost can be recouped and excessive duplication avoided. The industry may also consider new options such as pooling the cost of new infrastructure investment.

The above issues will preoccupy the regulatory authority in the coming years. The policies in the future to deal with these issues will have significant impact on the competitiveness of the Malaysian economy. Malaysia's desire to become a developed nation by 2020 will crucially depend on how well it nurtures the development of the telecommunications sector.

ACKNOWLEDGEMENT

The author gratefully acknowledges research support from the Centre on Regulation and Competition at the Institute for Development Policy and Management, University of Manchester. The author would like to thank K.S. Jomo, Moazzem Hossain, Harold Read, and Wan Faizal Wan Hassan for useful comments and suggestions. The usual caveat applies. This chapter was published earlier in the annals of *Public and Cooperative Economics*, **73**, pp. 521–38, 2002, Blackwell Publishing, Oxford UK.

NOTES

1. Adam and William (1995, p. 11), for example, regarded privatisation as the second phase in the NEP where assets accumulated by the government on behalf of the Bumiputeras were redistributed to individual Bumiputeras and Bumiputera institutions.
2. The detailed breakdown is as follows: Khazanah Nasional Berhad (36.1 per cent), Minister of Finance, Inc. (21.3 per cent), Bank Negara Malaysia (8.23 per cent) and Permodalan Nasional Berhad (4.5 per cent).
3. MIMOS or the Malaysian Institute of Microelectronic Systems, was established in 1985 and corporatised in 1996.
4. Equal Access was a policy formulated by the Minister and CMC's predecessor, JTM. Efforts to implement it was deferred from 1995 to 1999.
5. Equal Access by pre-selection was originally scheduled to implement on 1 January 2001. A review of this proposal will be made in mid-2001.
6. See http://www.britain.org.my/trade/sector summary/telecommunications.htm. Handset subsidies are also eroding the margin from cellular service business. See *Malaysian Business*, 1 March 2000.
7. In May 2001, British Telecom announced that it would be divesting its stake in Maxis.

ABBREVIATIONS

APRU	average revenue per unit
CMA	Communications and Multimedia Act
CMC	Communications and Multimedia Commission
CMCA	Communications and Multimedia Commission Act
IDD	international direct dialling
ISP	internet service provider
JTM	Jabatan Telekom Malaysia
MAS	Malaysian Airline System
MECM	Ministry of Energy, Communications and Multimedia
METP	Ministry of Energy, Telecommunications and Posts
NEP	National Economic Plan
NTP	National Telecommunications Policy
TM	Telekom Malaysia
RM	Ringgit Malaysia (Malaysian currency)
TMB	Telecom Malaysia Berhad
TNB	Tenga Nasional Berhad
STD	subscribers trunk dialling
STM	Syarikat Telekom Malaysia
USF	Universal Service Fund

USO	universal service obligations
VAN	value added network service

REFERENCES

Adam, C. and C. William (1995), 'Early Privatizations', in K.S. Jomo (ed.), *Privatizing Malaysia: Rents, Rhetoric, Realities*, Boulder, CO: Westview Press, pp. 11–14.

Business Times (2001), 9 May, Kuala Lumpur.

Communications and Multimedia Act 1998 (1998), Kuala Lumpur: Government of Malaysia.

Communications and Multimedia Commission (2000a), 'Guideline on substantial lessening of competition', Discussion Paper No. RG/SLC/1/00(1), Kuala Lumpur.

Communications and Multimedia Commission (2000b), 'Guideline on dominant position in a communications market', Discussion Paper No. RG/DP/1/00(1), Kuala Lumpur.

Communications and Multimedia Commission (2000c), 'Process for assessing allegations of anti-competitive conduct: An information paper', Information Paper No. IP/Competition/1/00(1), Kuala Lumpur.

Communications and Multimedia Commission (2000d), 'Access list determination and statement on access pricing principles', Consultation Paper, December, Kuala Lumpur.

Communications and Multimedia Commission (2001a), 'The regulatory framework for the communications and multimedia sector in Malaysia', mimeo, Kuala Lumpur.

Communications and Multimedia Commission (2001b), 'Report on a public inquiry under Section 55 of the Communications and Multimedia Act, 1998 on a system of universal service provision', 12 March, Kuala Lumpur.

Communications and Multimedia Commission (2001c), 'Commission determination on universal service provision, Determination No. 2 of 2001', Kuala Lumpur.

Department of Statistics (1999a), *State/District Data Bank*, Kuala Lumpur: Government of Malaysia.

Department of Statistics (1999b), *Social Statistics Bulletin*, Kuala Lumpur: Government of Malaysia.

Gomez, E.T. and K.S. Jomo (1999), *Malaysia's Political Economy: Politics, Patronage and Profits*, Cambridge: Cambridge University Press.

Isahak, D. (1989), 'Meeting the challenges of privatization in Malaysia', in W. Bjorn, P. Stern, T. Nulty and R. Stern (eds), *Restructuring and Managing the Telecommunications Sector*, Washington DC: World Bank, pp. 118–21.

Kennedy, L. (1990), 'Privatization and its policy antecedents in Malaysian telecommunications', Ph.D. Dissertation, Ohio University.

Kennedy, L. (1995), 'Telecommunications', in K.S. Jomo (1995), *Privatizing Malaysia: Rents, Rhetoric, Realities*, Boulder, CO: Westview Press, pp. 219–35.

Lee, C. (2000), 'Regulatory Reform in the Infrastructure Sector: The Malaysian Experience', mimeo, Kuala Lumpur.

Malaysian Business (2000), 1 March, Kuala Lumpur.

Malaysia Plans (sixth, seventh and eighth), Kuala Lumpur: Government of Malaysia.

Malaysia Plans (sixth and seventh), *Mid-term Review*, Kuala Lumpur: Government of Malaysia.

Naidu, G. and C. Lee (1997), 'The transition to privatization: Malaysia', in A. Mody (ed.), *Infrastructure Strategies in East Asia: The Untold Story*, Washington DC: World Bank, pp. 27–49.

National Telecommunications Policy (1994), Kuala Lumpur: Ministry of Energy, Telecommunications & Posts, Government of Malaysia.

Poon, W-C. (2000), 'Regulatory reforms in the Malaysian telecommunications sector', Master of Economics research paper, University of Malaya.

Syed, H.M. (1999), 'Communications and Multimedia Regulation: An Overview', mimeo, Kuala Lumpur.

Telekom Malaysia Berhad (2000), *1999 Financial Review*, Kuala Lumpur.

TIME dotCom (2001), Public Issue Prospectus, January.

www.ktkm.gov.my

www.britain.org.my/trade/sector summary/telecommunications.htm

6. The Philippines: Institutionalising a liberal telecommunications regime

Rhoda U. Reyes and Ramonette B. Serafica

After more than 300 years of rule by Spain, the Philippines came under United States control in 1898. Although the nation achieved independence from the US in 1946, the structure and organisation of the Philippine administration and economy remain to this day strongly influenced by the US model of capitalism. In telecommunications the Philippines adopted a predominantly privately-owned system from the outset. This followed the example of the US and so differed from the models of almost all other countries where telecommunications was originally publicly-owned and controlled. For more than 60 years, telecommunications in the Philippines was virtually a privately-owned monopoly with services provided by the Philippine Long Distance Telephone Company (PLDT). Reform of the industry had to await the overthrow of the discredited regime of President Ferdinand Marcos in 1986.

The primary objectives of this chapter are to explain the nature of the post-Marcos reforms of telecommunications and to examine the effects of these reforms upon the industry. Section 1 sketches the system of telecommunications in the Philippines during the pre-reform period; section 2 describes the reforms introduced during the late 1980s and early 1990s; section 3 outlines the regulation of the reformed telecommunications system; section 4 examines the performance of the industry during the late 1980s and 1990s, and provides a critique of the reforms; and finally, section 5 considers some lessons from the Philippine experience of telecommunications reform.

1. PHILIPPINE TELECOMMUNICATIONS BEFORE REFORM

The development of Philippine telecommunications commenced in 1872 with the laying of a telegraph line between Manila and the island of Corregidor. The first overseas submarine telegraph cable linking the Philippines with the

British Crown Colony of Hong Kong was laid in 1880. The Philippine telephone system was inaugurated in 1890, and in 1928 the country's four exchanges were taken over by the PLDT when this company was awarded a 50-year congressional franchise to provide nation-wide telecommunications services. PLDT was owned by the British Columbia Telephone Company but operated under American management. To complement rather than compete with PLDT, the Bureau of Telecommunications (BUTEL) was established in 1947 to provide and operate a small publicly-owned telephone network to service a number of rural areas (Telecommunications Office 1997).

In 1956 the US utility company, General Telephone and Electronics, bought PLDT from British Columbia Telephone. In 1967, however, the US company was forced to divest its ownership of PLDT due to 'nationalist pressures, domestic business demands, and economic pragmatism' (Benedicto-Wolf and Sussman 1995, p. 210). PLDT was acquired by Ramon Cojuangco, a member of the Philippine land-owning elite and friend and political ally of Marcos.

Although by the 1970s there were several minor telecommunications service providers throughout the country, PLDT was the only carrier with a national franchise and had no effective competition (Aquino 1994). Network development remained sluggish with the teledensity rate at less than 1 per cent. The telecommunications industry was widely considered to be highly inefficient, and had developed in a highly segmented manner in terms of service, technology and geography (Serafica 1998). Government regulation of the industry was limited and focused primarily on rate structuring (Dans 1985).

During the period of martial law from 1972 to 1981 the dominant position of PLDT was strengthened further. In 1973, Marcos issued a presidential decree instituting a subscriber investment plan which made the provision of PLDT services conditional upon subscribers purchasing a minimum of 180 non-voting shares in the company at par value of 10 pesos (P10) per share (about US$1.50 at the 1973 exchange rate) (Esfahani 1996)[1]. Although the decree's stated purpose was to widen public ownership in PLDT, its primary intention was to provide PLDT with a more suitable shareholding structure to facilitate fund raising on the international capital market (Manapat 1993; see also Coronel 1998 for a detailed account of how PLDT was able to maintain its dominant position during the Marcos, Aquino and Ramos presidencies).

In June 1973, five years before the expiration of PLDT's franchise, a study group from the World Bank visited the Philippines to evaluate the country's telecommunications infrastructure. The group reported that there was poor network design, duplication of facilities and widely varying standards of service throughout the country, and recommended the consolidation and interconnection of the various local telecommunications facilities.

In response, Marcos ordered the acquisition by PLDT of a number of private telecommunications companies, most notably the Republic Telephone Company, which was the country's second largest carrier. These acquisitions left PLDT with around 90 per cent of the Philippine telecommunications market (Sussman 1987, 1991a). All other service providers were required to interconnect with PLDT, whose franchise was extended for 25 years to 2003.

The expiration of the Laurel/Langley Agreement in 1974 ended the preferential rights of US shareholders to fully own and control Philippine public utilities. New arrangements applying to telecommunications facilities specified a minimum 60 per cent Filipino ownership and a maximum foreign ownership of 40 per cent. In 1979 the government formed as its policy-making body the Ministry of Transportation and Communication, now the Department of Transportation and Communications (DOTC). It also established the regulatory agency, the National Telecommunications Commission (NTC), and renamed BUTEL as the Telecommunications Office.

2. PHILIPPINE TELECOMMUNICATIONS REFORM

In the late 1970s and early 1980s a number of forces came together to provide both the context and rationale for reform of telecommunications in the Philippines. With widespread public sector inefficiency and the collapse of 'the state as entrepreneur' in a number of developing countries, multilateral lending institutions including the International Monetary Fund (IMF) and the World Bank pressed strongly for economic reforms. These reforms featured the privatisation of publicly-owned and operated services, and sectoral deregulation and liberalisation, which the lending institutions claimed would solve the problems of inefficient state enterprises and poor economic performance (Wellenius and Stern 1994; Petrazzini 1995).

In telecommunications in particular, rapid technological change and falling prices of equipment eroded the natural monopoly characteristics of the industry. This signalled the shift from monopoly to competitive provision, especially in light of increased globalisation of corporate activities (Noam and Wolfson 1997). These developments were complemented by public demand for more efficient and diverse telecommunications services. The possibility of addressing this demand by allowing alternative service providers to enter the market increased the pressure upon governments to introduce competition as a means to improve the availability and quality of telecommunications services. There was also increasing evidence from both developed and developing nations that telecommunications reform paved the way for increased foreign investment, expansion of telephone services and improvements in overall economic productivity (Saunders et al. 1994).

A major impetus for political and economic reform in the Philippines came in February 1986 when the corrupt Marcos regime was suddenly overthrown and Corazon Aquino became President. Prior to Marcos being deposed, the Philippines had experienced a burgeoning debt crisis, leaving the new administration with a national debt of around US$28 billion. The new economic policy settings adopted by Aquino recognised the shortage of domestic and foreign capital to finance infrastructure development, and yielded to pressure from the IMF, the World Bank and the US to open the Philippine economy to foreign investment.

One of Aquino's first decisions on assuming power was to form a Presidential Commission on Good Government (PCGG), which was charged with the task of tracing and recovering the wealth that Marcos had illegally accumulated during his time as president. One of the trails to that wealth led to PLDT, and the PCGG sequestered the company early in its investigation. However, the sequestration order was lifted after three months and no funds were ever recovered from PLDT (Sussman 1991b). The PCGG's failure to recover monies from PLDT was later described as indicative of the Aquino administration's inability 'to shield itself from the pressures – and temptations – brought by those who had everything to gain by keeping things as they were' (Coronel 1998, p. 125). It should be noted here that Cory Aquino is a member of the Cojuangco family and, as such, is related to the owners of PLDT.

Although unsuccessful in recovering funds from PLDT, Aquino can be credited for commencing the process of telecommunications reform. In 1987 an updated Philippine Constitution gave full recognition to the vital role of telecommunications in nation building. The following year Aquino appointed a National Telecommunications Development Committee resulting in the formation of a National Telecommunications Development Plan, which set out goals for the future growth of the industry. Further, Aquino issued the country's Investment Priorities Plan that included telecommunications for the first time as one of the priority areas for investment (PDCP 1988).

In 1989 two new companies, Eastern Telecoms and Philcom, were allowed entry into the international long-distance market, and Extelcom, the first cellular company to operate in the Philippines, was granted interconnection with PLDT after prolonged proceedings in the Supreme Court. A Municipal Telephone Projects Office (MTPO) was also created to spur the establishment of public calling offices (PCOs) in towns and municipalities throughout the country.

Notwithstanding these steps towards liberalisation, by 1990 PLDT was still the dominant operator in the industry. The quality of service remained poor, availability of service between urban and rural areas was widely uneven, and there was a large unmet demand for telephone connections (DOTC 1993). In 1991, PLDT's franchise was extended for another 25 years to 2028. This extension of the franchise was granted towards the end of the Aquino administration, and 12 years before the scheduled time of its expiration.

In July 1992 Aquino was succeeded by President Fidel Ramos who had an optimistic vision of the Philippines becoming a newly industrialised country by the twenty-first century. To this end he paved the way for further liberalisation of the telecommunications sector. In February 1993 Ramos issued an executive order (EO 59) requiring compulsory interconnection of all authorised public telecommunications operators to create 'a universally accessible and fully integrated nation-wide telecommunications network and thereby encourage greater private sector investment in telecommunications' (Office of the President 1993a, p. 1). Five months later he issued another executive order (EO 109) that called for the improvement of the local exchange service in urban and rural areas by adopting the Service Area Scheme (SAS) (Office of the President 1993b). The objectives of the SAS reform programme were to promote the development of telecommunications in the Philippines by expanding network coverage and the teledensity rate, and to provide the conditions for competition between PLDT and new service providers.

The framework for implementing the SAS was developed by the NTC following an exhaustive study and detailed consultations with major national and foreign telecommunications companies. The principle adopted was to grant new licences for international gateway facility (IGF) and/or cellular mobile telephone system (CMTS) services to nine private companies, and in turn require each of them to provide a local exchange carrier (LEC) service (that is, fixed-line) in a specified area. Each service area was considered to be sufficiently large in terms of population to enable the companies to be financially viable. The National Capital Region (Metro Manila), accounting for around 47 per cent of the country's installed lines, was divided into four separate service areas which were each paired with a more sparsely populated (and therefore less financially attractive) region of the country. With PLDT already having a nation-wide franchise, the new licences under the SAS reforms created a duopoly situation in Metro Manila with PLDT and the new carrier, and a triopoly in some of the provinces with PLDT, the new carrier and an existing small private firm operating in competition.

Network expansion was to be achieved by requiring licensee companies to install new local exchange lines. Each IGF licensee was required to install a minimum of 300 000 lines within three years of the date of their provisional

Table 6.1 Service area scheme, 1995

Licensee company	Local owners	Foreign partners	Service obligation (number of lines)	Service area
Smart	Vea-Fernando family Metro Pacific	First Pacific HK NTT (Japan)	IGF: 300 000 CMTS: 400 000	Regions 1, 3 CAR, NCR-D
Eastern Telecoms	Nieto-Africa family Govt of the Philippines	Cable & Wireless HK	IGF: 300 000	Region 2 NCR-A
PT&T	Santiago family RETELCOM	Korea Telecom	IGF: 300 000	Region 4A
Globe Telecom	Ayala family Ayala Corp	Singapore Telecom	IGF: 300 000 CMTS: 400 000	Regions 4B, 12 NCR-C
BayanTel	Lopez family Benpres Corp	Bell Atlantic (formerly US Nynex) Asia Infrastructure Fund Group Chase Capital Partners	IGF: 300 000	Region 5 NCR-B
Islacom	Delgado family Asiacom Networks	Shinawatra (Thailand) Deutsche Telekom	IGF: 300 000 CMTS: 400 000	Regions 6, 7, 8
Piltel/ Philcom	PLDT/ Enrile family	Comsat (US)	CMTS: 400 000 IGF: 300 000	Regions 9, 10, 11
Digitel	Gokongwei family JG Summit Corp	AB Telia (Sweden)	IGF: 300 000	Regions 1–5 (Luzon-wide)

Notes: CAR: Cordillera Administrative Region (located north of the Philippines); NCR: National Capital Region: divided into 4 sub-sections, A-D, representing different parts of Manila; Regions: 1. Ilocos; 2. Cagayan; 3. Central Luzon; 4A. Southern Tagalog (east); 4B. Southern Tagalog (west); 5. Bicol; 6. Western Visayas; 7. Central Visayas; 8. Eastern Visayas; 9. Western Mindanao; 10. Northern Mindanao; 11. Southern Mindanao; 12. Central Mindanao

Sources: NTC (1997–98); Company and industry reports

authority (PA) to install, operate and maintain a local exchange service. For every 10 lines installed in an urban area, one line was to be installed in a rural area. Each CMTS licensee was required to install 400 000 lines, also within three years from the date of its PA. All IGF and CMTS licensees were required to deposit an escrow amounting to 20 per cent of their projected investment for the first two years of the project, and to post a performance bond equivalent to 10 per cent of their first two years' investment. This amount was to be forfeited in favour of the government in the event of the carrier failing to comply with the conditions of its PA (Office of the President 1993b).

Table 6.1 sets out the names of the nine companies licensed under the SAS scheme together with their local and foreign equity owners, their respective service areas and the service obligations of their licences. (The data in the table relate to the commencement of the SAS scheme in 1995 and have changed somewhat since that time with the modification and progress of the scheme.) Reflecting the foreign ownership limitations upon Philippine telecommunications, the shareholding in the SAS licensee companies typically comprises 60 per cent local ownership and a maximum 40 per cent equity holding by a foreign partner. Domestic ownership in each case comprises a local elite family together with a Philippine company or government authority. BayanTel, for example, is owned by the Lopez Group led by Benpres Holdings Corporation with a 68.76 per cent equity holding, while the foreign partners comprise Bell Atlantic with 20.01 per cent, Asia Infrastructure Fund Group (10.3 per cent) and Chase Capital Partners (0.93 per cent).

3. REGULATORY FRAMEWORK

In February 1995 the Philippine Congress passed the Republic Act (RA 7925, hereafter 'the Act') that was designed to direct the future development of telecommunications in the country (Congress of the Philippines 1995; see also Coronel 1998 for an account of the legislative process leading to the adoption of RA 7925). The Act incorporated executive orders 59 and 109, restated the constitutional provisions recognising telecommunications as vital to economic advancement, and reaffirmed that the private sector was to take on the task of facilitating the development of the industry. Most importantly, the Act sets out the regulatory framework for the industry. Its salient provisions are as follows:

- The Philippine government is to promote a competitive environment allowing carriers to provide telecommunications services while ensuring their financial viability.

- Telecommunications policy development is the responsibility of the DOTC. The department is charged with formulating the government's plans for the industry, coordinating research, and negotiating international telecommunications agreements.
- The NTC is the industry regulator under the administration of the DOTC. The NTC is to be independent in its regulatory functions with its decisions appealable only to the Supreme Court. It is headed by a Commissioner with two Deputy Commissioners.
- The NTC is responsible for the granting of new licences for providing and operating telecommunications services.
- The NTC determines the prices operators charge their subscribers.
- The NTC is to 'mandate a fair and reasonable interconnection of facilities of authorised public network operators and other providers of telecommunications services' (Congress of the Philippines 1995, Article III5c). Access charges and revenue sharing arrangements between interconnecting carriers are to be negotiated between parties and all agreements between parties are to be endorsed by the NTC. Where the parties fail to reach agreement, the dispute is to be submitted to the NTC for resolution.
- Telecommunications consumers have the right to be provided with a telephone connection within two months of applying. They also have the right to non-discriminatory provision of facilities and to regular, timely and accurate billing of services. User and consumer groups are allowed to participate in public hearings and private consultations on issues and concerns relevant to telecommunications policy development.

In February 1997 the Philippines became one of 70 signatories to the World Trade Organization (WTO) negotiations on basic telecommunications services, reasserting its commitment to liberalise and promote competition in the sector. The Philippine government also signed the General Agreement on Trade in Services thus committing the nation, among other things, to opening up the telecommunications market to foreign investment (within the Constitutional limitation of 60 per cent minimum domestic ownership), equal treatment of national and foreign telecommunications operators, and independent, stable and transparent regulation of the industry (WTO 1997).

4. EFFECTS AND CRITIQUE OF REFORMS

The liberalisation of Philippine telecommunications by the Ramos administration had the effect of stimulating a moribund industry that is now one of the growth sectors of the economy. In 1998, while industrial production declined by 1.7 per cent, and agricultural production by 6.6 per cent, the service sector recorded a growth rate of 3.5 per cent. The National Economic and Development Authority (NEDA), which is the country's economic planning body, attributed the service sector's performance largely to growth in the telecommunications sector (NEDA 1998; *Business World* 1998). Whereas telecommunications contributed 1.68 per cent to GDP in 1995, the sector accounted for 11 per cent of GDP during the second quarter of 2000, credited mainly to the increased demand for mobile communications (PDCP 1996; Maguilas 2000). Opening up telecommunications to more market players increased investment in the industry from P90.2 billion (US$3.5 billion) in 1992 to P304.7 billion (US$10.3 billion) in 1997 (NTC 1998 Annual Report). Total market size for Philippine telecommunications in 1997 was estimated at US$1.4 billion (US International Trade Commission 2000).

Other benefits of liberalisation were seen in higher teledensity, greater consumer choice of telecommunications services, and lower prices for equipment and services. The number of lines was increased substantially from 0.74 million in 1992 to 6.6 million in 1998 (NTC 1997–98). Figure 6.1 depicts the slow growth of the telecommunications sector during the 1980s and early 1990s, followed by a steep rise (albeit from a low base) from 1993 with the implementation of the liberalisation programme.

In addition to voice telephony, various value-added services are now available in the Philippines, including caller identification, call blocking, call waiting, call divert, teleconferencing, text-messaging and three-way networking. Prices of telecommunications consumer equipment have declined significantly. For example, one report traced the price of cellular mobile phones from P40,000 (approximately US$1,500) in 1991, to P15,000 (US$550) in 1993, and to P2300 (US$90) in 1996 (Tiglao 1996). The prices of international calls fell by similar proportions during the same period.

Another beneficial effect of the Ramos reforms has been the improved performance of PLDT in response to increased market competition. In 1993 PLDT launched the Zero Backlog Programme whereby it committed itself to install, by 1998, 2 million new lines. This was four times the number it had installed over the previous 60 years! The company substantially reduced the installation time of telephone lines and achieved its target under the programme. It increased its line capacity from 0.5 million in 1993 to 2.4 million in 1998. Of these 2.4 million lines, about 1.6 million were subscribed lines, representing a capacity utilisation of around two-thirds (Morgan Stanley Dean Witter 1999).

Under the SAS scheme the nine licensee companies had an obligation to install 4 million new telephone lines (see Table 6.1). An additional 1.3 million lines were required in the revised rollout plan. By 1999, when all carriers were scheduled to have completed their rollouts, around 4.6 million lines had been added, somewhat less than the required 5.3 million. Five of the licensees fell short of their committed number of lines: Islacom, Piltel, Philcom, Philippine Telephone and Telegraph (PT&T) and Eastern Telecoms. The NTC had power to impose sanctions upon these companies by revoking their licences and forfeiting their performance bonds. In the event, however, the NTC decided not to impose sanctions on the companies, reasoning that cancellation of the licences would be detrimental to the interests of these companies' subscribers (Fajardo 1999a). Instead, the NTC decided to reward the carriers that did deliver on their landline obligations within the scheduled time by authorising them to expand their service areas (for example, BayanTel). This decision led to some areas being highly competitive. For example, in parts of Southern Luzon, Digitel, PT&T and PLDT all compete for a share of the local market.

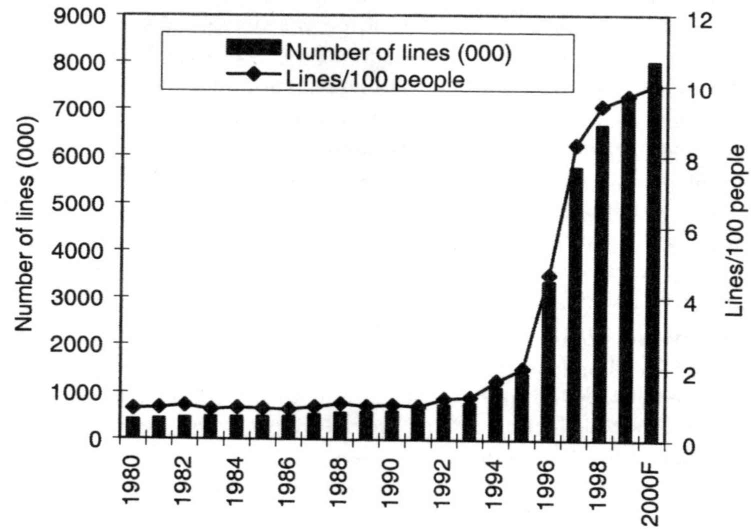

Note: F = forecast

Sources: NTC Annual Reports; Industry reports

Figure 6.1 Philippine telephone line and teledensity growth, 1980–2000

Notwithstanding the increased competition brought about by the SAS reforms, the incumbent, PLDT, remains the dominant player in the local exchange market. PLDT is the only company that is allowed by the NTC to provide a nation-wide service, all other firms being confined to their respective service areas. It can be seen from Figure 6.2 that in 2000 PLDT accounted for 56 per cent of the LEC market, with no other single carrier having more than 11 per cent market share.

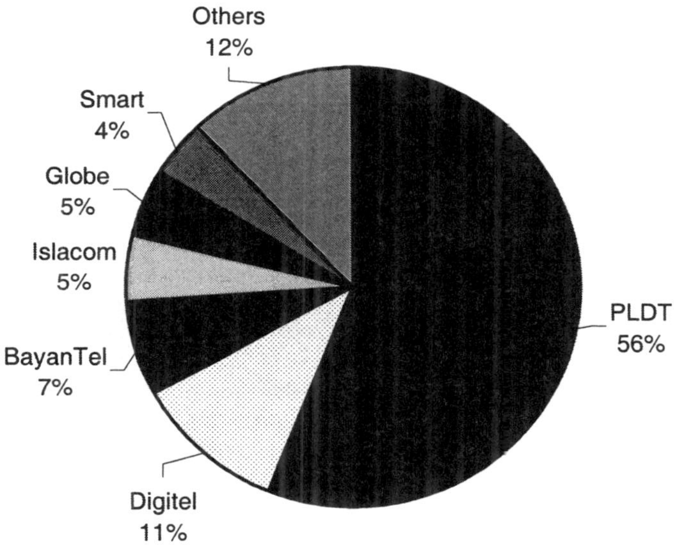

Notes: Data are for period January–December 2000
'Others' represent more than 60 small private telecommunications service providers each vying for a niche in its local market

Source: NTC (2001a)

Figure 6.2 LEC market share, 2000

Figure 6.3 shows that the distribution of the cellular mobile market differs considerably from that of the fixed-line market. In 2000, two players in the sector accounted for 77 per cent of total revenue – Smart (39 per cent) and Globe (38 per cent) – while the remaining 23 per cent was shared among the three other mobile operators.

A study conducted by the DOTC in 1998 on the state of access to telecommunications services in the Philippines revealed that 'installed teledensity' – that is, the proportion of the residential population with a fixed telephone line passing their homes – was approximately 8 per cent. However, less than half of the available capacity was actually subscribed, giving an actual teledensity of subscribed lines of 3.8 per cent, including business connections. For residential subscribers only, the teledensity rate is even lower at 2.77 per cent (DOTC 2000). These data suggest that lack of service supply is no longer the major reason for the low teledensity level in the Philippines. Rather, the main barrier to access to telecommunications by households is affordability.

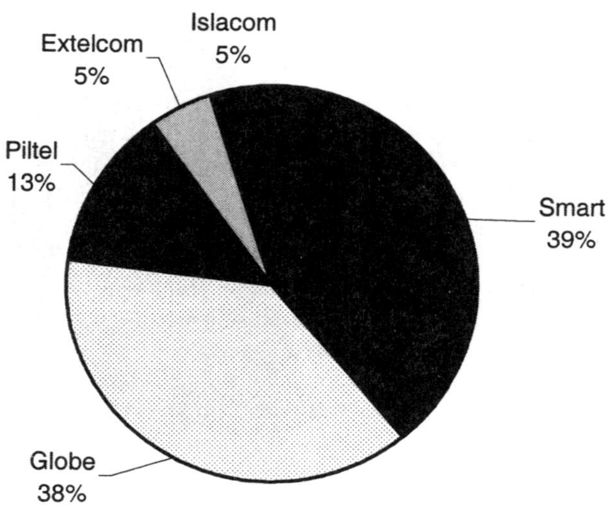

Note: Data are for period January–May 2000

Source: NTC (2001b)

Figure 6.3 CMTS market share, 2000

This finding is consistent with an independent unpublished study conducted by the authors of this chapter. In a survey of selected urban and rural areas it was found that affordability remains a major problem for those households whose monthly income is less than P10,000 (approximately US$250 at the December 1998 exchange rate). With a minimum wage of only P200 (US$5) per day, and the cost of food accounting for nearly half of that, subscription to a telephone service is a luxury few can afford. For most Filipino households the basic needs of food, shelter, clothing and education take priority over telecommunications.

For Filipinos who can afford to subscribe, reliability of the telephone service continues to be a problem with call failure rates well in excess of NTC allowable levels. The main problem concerns interconnection. The Act provides for interconnection to be negotiated among carriers and, if the parties fail to reach agreement within 90 days, for the NTC to intervene. The NTC has developed a template regarding the joint sharing, between access providers and access seekers, of the costs of interconnection facilities and their maintenance. However, it does not have the administrative capacity to enforce its determinations. The Act lacks specificity regarding the role and procedures of the regulator in relation to interconnection, and does not give the regulator power to impose sanctions on delinquent carriers. This situation allows PLDT in particular, as the dominant carrier, to delay the completion of interconnection agreements and otherwise avoid its interconnection obligations. PLDT has not been the only culprit in this area. Other carriers have unilaterally discontinued existing interconnection agreements (for example, BayanTel's agreement with Smart) and the NTC has been powerless to act (Fajardo 1999b).

As mentioned in the previous section, the prices of telecommunications services in the Philippines are determined by the NTC. The only exception to this relates to the prices for mobile handsets, which are market determined. Competition in Philippine telecommunications is thus mainly confined to the range and quality of services (non-price features) among the different carriers. Most of the telecommunications operators have incurred loans denominated in US dollars from international finance corporations to fund their establishment and development in the Philippine market. Because the NTC is required to ensure the financial viability of these carriers, the fluctuation in the Peso–US dollar exchange rate is a major factor taken into account by the NTC in its price determination process.

Another development in the post-reform period has been a move towards consolidated ownership among telecommunications carriers, most notably in November 1998. First Pacific Hong Kong, which already held a 40 per cent equity interest in Smart (one of the two major operators in the mobile market) acquired a 17.2 per cent interest in PLDT (the dominant fixed-line carrier). Forging a common ownership link between the major players in the fixed-line and mobile markets will generate economies of scale and scope between the two companies, but it also has the potential to stifle competition in the industry. It is also likely to cause other smaller players in both the fixed and mobile markets to consider consolidating their resources in an attempt to provide a degree of countervailing power to the combined might of PLDT and Smart.

It was mentioned in section 2 that the Aquino administration established the MTPO for the purpose of setting up public calling offices (PCOs) throughout the country. Approximately 1300 PCOs were planned to be built during the 1990s. In the event, some 1000 PCOs were installed: around 700 by the MTPO, and 300 by private carriers. However, many of these were not adequately maintained. The MTPO was dissolved in 1999 and the government announced that it planned eventually to privatise all government telecommunication facilities (MTPO 1999).

Following the passage of the 1996 Telecommunications Act in the US, which allows for the common ownership of telecommunications and broadcasting services, a Draft Convergence Bill with similar provisions was filed in the Philippine Congress. However, RA 7925 currently prohibits cross-ownership between telecommunications and broadcasting, and the Philippine Constitution provides that all broadcasting services must be 100 per cent owned by Filipinos, while telecommunications services may have up to 40 per cent foreign equity ownership. The adoption of the Convergence Bill by the Philippine Congress will therefore require prior amendment to the Constitution.

5. LESSONS FROM THE PHILIPPINE EXPERIENCE

Although the Philippines has made some bold moves to liberalise telecommunications there remain intractable problems and much to be done to enhance competition in the industry. The decision to demonopolise the industry was certainly a correct one. Liberalisation has had the desirable effects of expanding the availability of telecommunications services, increasing consumer choice, improving quality of service and lowering prices. Equally important, it has stimulated the telecommunications industry, which is now one of the most dynamic sectors of the economy.

A major weakness in Philippine telecommunications, however, concerns its regulation. The NTC lacks the resources and executive power to properly regulate the industry. In particular, it does not possess sufficient authority to moderate the market power held by PLDT. The telecommunications legislation does not provide the NTC with sufficiently strong and clear rules to enable it to ensure fair competition among market players or to respond to rapid developments in the industry. It should be noted, however, that certain issues regarding competition, for example, mergers and acquisitions, are beyond the regulatory responsibility of the NTC. Also, the requirement upon the NTC to determine nearly all prices operators can charge consumers severely restricts competition in the industry, and needs, in time, to be phased out to allow effective competition to develop.

The Philippine experience demonstrates that simply removing restrictions to entry into the telecommunications market will not by itself create a truly competitive environment. The dominant incumbent operator holds a strategic advantage that it can and does use to minimise competition. PLDT has been particularly adept in delaying or limiting interconnection by competitors with its network, and in resisting attempts by the NTC to impose terms and conditions of access necessary for effective competition.

While one of the main objectives of the telecommunications reforms was to reduce the market power of PLDT by licensing a number of rival companies, this objective has been at least partly thwarted by the formation in 1998 of the common ownership linkage between PLDT and Smart. This consolidation of ownership has the potential to stifle competition. In light of the trend towards convergence in the communications industries, a clear and effective policy on mergers and acquisitions should be considered an urgent item on the reform agenda (Serafica 2000).

The ownership consolidation between PLDT and Smart also points to an inherent defect in the SAS scheme for telecommunications reform. While the SAS creates competition for PLDT, the competition is uneven. PLDT remains the only carrier that is allowed by the NTC to provide a nation-wide service and, as such, benefits from economies of scale not enjoyed by its smaller and geographically segmented rivals. Together with the advantage of PLDT being the incumbent with over 70 years' experience in the Philippine telecommunications market, it is extremely difficult for the new carriers to challenge PLDT's market dominance. This suggests that consideration should be given to encouraging ownership consolidation among the SAS licensee companies, and/or to the granting of a second national telecommunications franchise.

The history and political economy of the Philippines have together shaped the current state of the country's telecommunications. Given the defects in the reform scheme, problems of governance in the industry, lack of an effective competition policy, and the system of patronage politics that permeates the country, it may be some time before the goal of a fully liberalised and genuinely competitive telecommunications environment in the Philippines is realised.

ABBREVIATIONS

BUTEL	Bureau of Telecommunications
CMTS	cellular mobile telephone system
DOTC	Department of Transportation and Communications
IGF	international gateway facility
IMF	International Monetary Fund
LEC	local exchange carrier

MTPO	Municipal Telephone Projects Office
NCR	National Capital Region
NEDA	National Economic and Development Authority
NTC	National Telecommunications Commission
PA	provisional authority
PCGG	Presidential Commission on Good Government
PCOs	public calling offices
PDCP	Private Development Corporation of the Philippines
PLDT	Philippine Long Distance Telephone Company
PT&T	Philippine Telephone and Telegraph
SAS	Service Area Scheme
WTO	World Trade Organization

NOTE

1. Throughout this chapter the exchange rates used are those current at the time referred to.

REFERENCES

Aquino, T. (1994), 'The Philippines', in E. Noam, S. Komatsuzaki and D. Conn (eds), *Telecommunications in the Pacific Basin: An Evolutionary Approach*, Melbourne: Oxford University Press, pp. 177–200.

Benedicto-Wolf, Alexandrina and Gerald Sussman (1995), 'Privatization of telecommunications: lessons from the Philippines', in B. Mody, J.M. Bauer and J.D. Straubhaar (eds), *Telecommunications Politics: Ownership and Control of the Information Highway in Developing Countries*, Malwah, NJ: Lawrence Erlbaum Associates, pp. 201–24.

Business World (1998), 'Telecom sector boosts services for '98: NEDA', 24 December.

Congress of the Philippines (1995), *Republic Act 7925: An Act to Promote and Govern the Development of Philippine Telecommunications and the Delivery of Public Telecommunications Services,* 20 February, Congress of the Philippines, Manila.

Coronel, Sheila (1998), 'Monopoly', in S. Coronel (ed.), *Porks and Other Perks*, Manila, Philippines: Center for Investigative Journalism, pp. 113–49.

Dans, Jose P., Jr (1985), 'Philippine telecommunication policy development', in S. Rahim and D. Wedemeyer (eds), *Telecom Pacific,* Hawaii: Pacific Telecom Council, pp. 137–43.

DOTC (1993), National Telecommunications Development Plan 1991–2010: Executive Summary, July 1993 Update, Department of Transport and Communications, Manila.

DOTC (2000), *Universal Access Benchmark Study*, Department of Transport and Communications, Manila.

Esfahani, H.S. (1996), 'The political economy of the telecommunications sector in the Philippines', in B. Levy and P. Spiller (eds), *Regulations, Institutions and Commitment: Comparative Studies of Telecommunications*, New York: Cambridge University Press, pp. 145–203.

Fajardo, R.P. III (1999a), 'NTC won't cancel licenses of erring carriers – Santiago', *Philippine Telecom Digest*, **39**, August, 33.

Fajardo, R.P. III (1999b), 'Interconnection: one player too many?', *Philippine Telecom Digest*, **38**, July, 1–14.

Maguilas, A.L. (2000), 'Telecom seen spurring growth of RP economy', *Manila Times*, 9/5, 3B.

Manapat, R. (1993), *Wrong Number*, Madrid: Parwue del Buen Retiro.

Morgan, Stanley Dean Witter (1999), *Asia/Pacific Investment Research – Philippines: Telecommunications*, Hong Kong, Morgan Stanley Asia Limited, 7 April.

MTPO (1999), *MTPO Report: Linking the Countryside and the Nation to the World*, Quezon City: Municipal Telephone Projects Office.

NEDA (1998), 'Real gross national product by expenditure shares 1997–98', *Philippine Economic Indicators Online* (www.neda.gov.ph/TheEconomy/EconomicPerformance/4Q98.htm, accessed 6 July 2001).

Noam, E.M. and A.J. Wolfson (1997), 'The end of territoriality in communications', in E.M. Noam and A.J. Wolfson (eds), *Globalism and Localism in Telecommunications*, New York: Elsevier, pp. xvii–xxxix.

NTC (1997–98), *Annual Reports*, Manila: NTC.

NTC (2001a), 'Telephone distribution per operator as of December 2000', Consumer Information (http://www.ntc.gov.ph/consumer-frame.html, accessed 7 July 2001).

NTC (2001b), 'Cellular subscribers per operator as of May 2000', Consumer Information (http://www.ntc.gov.ph/consumer-frame.html, accessed 7 July 2001).

Office of the President (1993a), *Executive Order No. 59: Prescribing the Policy Guidelines for Compulsory Interconnection of Authorised Carriers in Order to Create a Universally Accessible and Fully Integrated Nationwide Telecommunications Network and Thereby Encourage Greater Private Sector Investment in Telecommunications*, 24 February.

Office of the President (1993b), *Executive Order 109: Policy to Improve the Provision of Local Exchange Service*, 12 July.

PDCP (1988), 'Future directions of the telecommunications industry', *Industry Digest*, Manila: PDCP, September–October, 1–8.

PDCP (1996), 'Telecommunications industry: competition benefits consumers – Parts 1 & 2', *PDCP Bank Industry Digest*, **20** (8/9), 1–10.

Petrazzini, B. (1995), *The Political Economy of Telecommunications Reform in Developing Countries: Privatization and Liberalization in Comparative Perspective*, Westport, CT: Praeger.

Saunders, R.J., J.J. Warford and B. Wellenius (1994), *Telecommunications and Economic Development*, 2nd edn, Baltimore, MD: Johns Hopkins University Press.

Serafica, R.B. (1998), 'Was PLDT a natural monopoly? an economic analysis of pre-reform Philippine telecoms,' *Telecommunications Policy*, **22** (4/5), 359–70.

Serafica, R.B. (2000), *Competition in Philippine Telecommunications: A Survey of the Critical Issues*, Discussion Paper No. 2000–15, Manila: APEC Study Center Network.

Sussman, G. (1987), 'Banking on telecommunications: the World Bank in the Philippines', *Journal of Communication*, **87**, 90–104.

Sussman, G. (1991a), 'Telecommunications for transnational integration: the World Bank in the Philippines', in G. Sussman and J. Lent (eds), *Transnational Communications: Wiring the World*, Thousand Oaks, CA: Sage, pp. 43–64.

Sussman, G. (1991b), 'The transnationalization of Philippine telecommunications: postcolonial continuities', in G. Sussman and J. Lent (eds), *Transnational Communications: Wiring the World*, Thousand Oaks, CA: Sage, pp. 125–49.

Telecommunications Office (1997), *Annual Report and 50th Anniversary Supplement*, Manila: TELOF.

Tiglao, R. (1996), 'Mayhem in the market', *Far Eastern Economic Review,* **159** (42), 17 October, 54, 56.

US International Trade Commission (2000), *Telecommunications Market Overview for the Philippines* (http://www.usitc.gov, accessed 23 February 2000).

Wellenius, B. and P. Stern (1994), 'Implementing reforms in the telecommunications sector: background, overview, and lessons', in Bjorn Wellenius and Peter Stern (eds), *Implementing Reforms in the Telecommunications Sector: Lessons from Experience*, Washington DC: World Bank, pp. 1–63.

WTO (1997), 'The Philippines: schedule of specific commitments – supplement 2', *Trade in Services,* GATS/SC/70/Suppl.2, 11 April (97–1386).

7. Thailand: The road to telecommunications liberalisation

Nittaya Soonthonsiripong

Historically the development of telecommunications in Thailand has been shaped profoundly by extensive state ownership. Developments in the global and domestic contexts of the telecommunications market in Thailand from the late 1980s in particular have forced Thai policy-makers to shift their philosophy on telecommunications as both a public service and a matter of national security. Today the Thai government is pursuing a programme of market liberalisation and encouraging increased competition that will improve both price and service for Thai consumers, and strengthen telecommunications as an industry of great importance to the development of the national economy.

This chapter examines the shift from state monopoly to market liberalisation. It discusses the state of the fixed-line and mobile phone markets that comprise the industry, and considers the reform process that is now underway. It describes steps that the government is taking to liberalise the market in a plan to proceed through corporatisation to privatisation. These steps centre on converting the concessions that the government established in the late 1980s and early 1990s, and reforming both legislation and the state-owned enterprises (SOEs) that have dominated the telecommunications industry. The chapter also considers the role of an independent regulator in the proposed privatised market. In conclusion, the chapter identifies the lessons that Thai policy-makers have learned as the telecommunications market proceeds to full liberalisation, scheduled for 2006.

1. STATE OWNERSHIP OF THAI TELECOMMUNICATIONS

1.1 Origins of State Monopoly

In the second half of the nineteenth century the governments of France and Britain, then both colonial powers, were in competition to lay telegraph cables in Thailand. The French sought from the Thai government exclusive rights to lay telegraph cables, seeking to block British influence in Indochina. Without specific telecommunications legislation in Thailand at that time, there were no restrictions on who could serve as telecommunications operators. Even then, however, the Thai government recognised foreign ownership in the nation's communications system as a concern for national security. The government thus circumvented the conflict between the colonial powers by authorising only a national body, the Post and Telegraph Department (PTD), to establish and operate the national telegraph system. The department installed the first telegraph line in 1875.

Telecommunications is one of three major sectors that comprise Thailand's communications industry, along with postal services and broadcasting. From the outset telecommunications was seen as a natural monopoly, and so the philosophy underlying the sector was that the state had a responsibility to protect the public interest by providing telecommunications services to its citizens at affordable prices. As Yoonaidharma (1993) has observed, since telecommunications was considered to be a public service, only the state was given the right to provide that service.

The administration of the telecommunications sector has been undertaken by four government organisations, namely, the Ministry of Transport and Communications (MOTC), the above-mentioned PTD, the Telephone Organisation of Thailand (TOT) and the Communications Authority of Thailand (CAT).

1.2 Legislation

Legislation concerning the telecommunications sector has helped shape all three of the communications sectors. The first telecommunications legislation came into effect in the 1930s, reserving the right of SOEs to provide and manage all communications services. Key pieces of legislation concerning the administration and operation of telecommunications have been the Telegraph and Telephone Act 1934, the Telephone Organisation of Thailand Act 1954 ('TOT Act') and the Communications Authority of Thailand Act 1976 ('CAT Act'). The TOT Act and the CAT Act established the TOT and CAT respectively. These acts have been central in determining the structure of the telecommunications market in Thailand by codifying the rights of the state to provide services.

The Thai Constitution reserves the right of the TOT to be the owner of the fixed-line networks and sole provider of fixed-line services (TDRI 1997). The TOT Act specifies the responsibilities of TOT, which include the provision of domestic long-distance services, regional calls (to Laos and Malaysia), and multi-access radio telephones. It also empowers the telecommunications SOEs to grant licences to and engage in joint ventures with private sector companies.

The CAT Act gives legal responsibility to CAT for two main types of communications services, namely, international calls and non-voice services (for example,, postal and telex). The Radio Broadcasting and Television Act 1955 reserves the exclusive right of the PTD to allocate radio spectrum frequencies and to manage wireless communications. In recent years the PTD has also been responsible for domestic satellite communications. The administrative structure for telecommunications in Thailand and the responsibilities of the PTD, TOT and CAT are summarised in Figure 7.1.

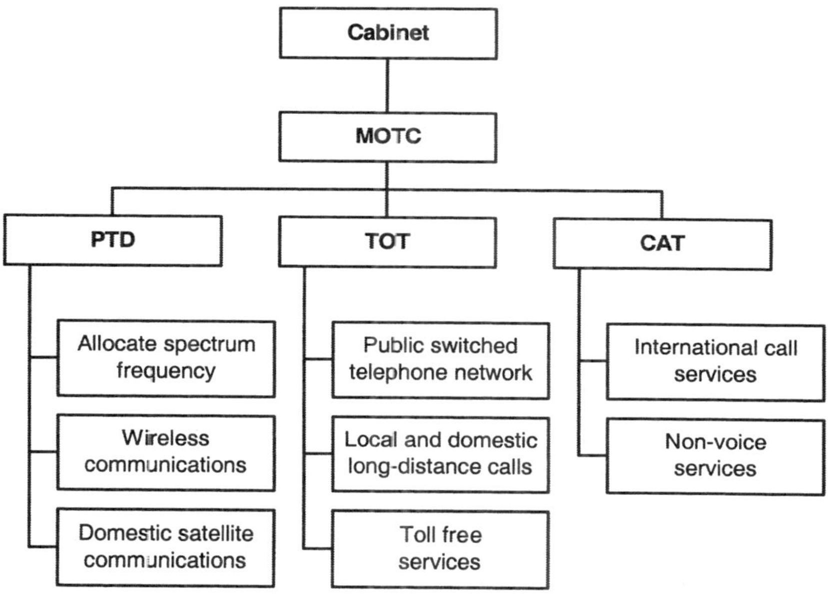

Figure 7.1 Structure of administration of Thai telecommunications

1.3 Price Setting

The TOT and CAT have been given regulatory powers relating to the pricing of telecommunications services under their respective acts. Price setting strategies, however, have to be approved by both the MOTC and Cabinet. All call charges are subject to regulation with the exception of mobile phone calls, which are market determined.

The TOT regulates the prices of voice services excluding international calls. The pricing system for fixed-line telephones provides that the same prices are charged to both business and residential subscribers, with tariffs in the form of access and usage charges. The access charges for a fixed line consist of a (first time only) installation fee, a refundable bond, and a monthly connection fee. The installation fee, however, is sometimes waived to attract new subscribers (see below).

The TOT also sets the prices for local and domestic long-distance calls. Subscribers are charged for local calls on a flat-rate basis. Long-distance calls are subject to time-zone metering, which provides for charging according to distance and length of calls. Peak and off-peak pricing applies to the basket of domestic long-distance calls, and tariffs are currently classified into six zones: 0–50; 51–100; 101–200; 201–350; 351–500; and over 500 kilometres. This pricing regime is of course inconsistent with the concept of the 'death of distance' in telecommunications, and is already beginning to give way to the trend towards a single rate for long-distance calls using VoIP (voice over internet protocol) technology.

In spite of the restriction of telecommunications prices being set by TOT and CAT, competition between operators emerges from time to time. For example, to attract new subscribers, operators regularly announce special package offers that include waivering installation fees, providing complimentary equipment such as facsimile machines, and/or reduced prices on personal communications telephone (PCT) handsets.

The prices for international calls are set by CAT. As with domestic long-distance calls, time-zone metering applies to international calls. All countries are classified into various zones for which different rates are charged. For example, Japan, Hong Kong and Cambodia comprise one such zone.

2. THE TELECOMMUNICATIONS SECTOR

In common with many developing countries, Thailand traditionally has had a poor telecommunications system. However, with advances in telecommunications technologies, and following the reforms of the early 1990s (see section 3), the industry has developed strongly in recent years.

2.1 Fixed-line Market

The PDT was initially charged with providing and operating fixed-line services. From 1954 this role was gradually transferred to TOT until, from 1969, TOT assumed full responsibility for these services (TOT 1991).

The telecommunications networks are bifurcated into the Bangkok Metropolitan Area (BMA) and the provincial telecommunications areas (PTAs). The BMA covers Bangkok and three peripheral provinces,

Pathumthani, Nontaburi and Samutprakarn. The PTAs comprise 72 provinces that are divided into five telecommunications zones.

Prior to regulatory reform in the early 1990s the number of people on the waiting lists for fixed lines was greater than the number of subscribers listed in Thailand's telephone books! About 3 million lines provided fixed-line services to the population of 58 million. By June 2001, however, the telecommunications system had a capacity of 7.8 million lines (for a population of 61.9 million). Three operators – TOT, TelecomAsia (TA) and Thai Telephone and Telecommunications (TT&T) – offered services in the fixed-line market. TOT and TA provided about 4.5 million lines in the BMA, while the TOT and TT&T provided around 3.3 million lines in the PTAs (TOT 2001).

Initially the TOT granted a 25-year concession to TA to install and operate 2 million fixed lines in the BMA, and a concession to TT&T for 1 million fixed lines in the PTAs. The TA project began operating in 1992, and TT&T started its project the following year. In 1995 the TOT amended its agreements with these two companies, adding 0.6 million lines to the TA agreement and 0.5 million lines to the agreement with TT&T. TA was thus responsible for installing 2.6 million lines in the BMA, and TT&T for installing 1.5 million lines in the PTAs.

The impact of the economic crisis in 1997 produced a surplus of supply, with sales in all line capacities lower than expected. By mid-2001 7.8 million lines were available nation-wide, but only three-quarters of line capacity was connected to subscribers. Although several packages were offered in an attempt to increase the number of subscribers – for example,, free installation, low rates for internet access, and a single rate for VoIP long-distance calls – only two-thirds of line capacity in the BMA was in use. The situation in the PTAs was somewhat better balanced, although here supply to many non-metropolitan communities did not meet demand. As a result, in mid-2001 the teledensity rate in the PTAs was approximately 5 per cent, whereas in the BMA it was around 32 per cent (TOT 2001).

2.2 Mobile Market

Cellular mobile telephones were introduced into Thailand in 1989. The TOT was the first operator in the field with the Nordic Mobile Telephone (NMT) system. Two other operators, Advanced Info Services (AIS) and Total Access Communications (TAC), have also been significant players. AIS, as delegated by TOT, uses the digital global system for mobile (GSM), while TAC has an agreement with CAT under which it offers the Personal Communications Network (PCN) system.

There are only a few operators in Thailand's mobile phone market, which is still relatively small. The high cost of both mobile handsets and calls has been responsible for the slow growth in this market. However, recent vigorous price competition in this sector, especially between TAC and AIS, is steadily increasing the number of cellular subscribers. By mid-2001 there were around 4.5 million mobile phone subscribers (up from 3.6 million in December 2000), representing a penetration rate of 7.26 per cent (TOT 2001).

Nevertheless, a new entrant, CP Orange (a joint venture between CP Telecom, TA and Orange UK), was due to commence cellular services in December 2001, although with a much smaller network than either AIS or TAC. In addition, discussions were underway between TOT and CAT regarding the possible introduction of a digital cellular system in the 1.8–1.9 GHz band with the view to increasing the number of cellular mobile subscribers by offering a low monthly charge (*Telecom Journal* 2001a).

3. TELECOMMUNICATIONS REFORM IN THAILAND

Until recent years the weakness of Thailand's telecommunications infrastructure hindered the nation's economic development. Particularly during the 1980s, many services in Thailand became inadequate to meet the increasing telecommunications demands of the Thai business community. Business users experienced problems in placing international calls, sending facsimile messages and receiving domestic telephone calls. These difficulties have been attributed to the shortcomings of the SOEs responsible for the nation's telecommunications system. The SOEs, in turn, had inadequate budgets, inefficient management and a deficiency of skilled labour (Soonthonsiripong 1998).

3.1 The BTO Regime

One response to the failure of SOEs to satisfy the telecommunications market effectively and efficiently was the introduction in 1986 of a system labelled BTO (Build–Transfer–Operate). Under the BTO scheme a private company was granted a concession to build a telecommunications facility, but immediately upon completion the concessionaire was required to transfer ownership of the facility to a SOE. The concessionaire was then permitted to operate the facility for a fixed period immediately after transferring ownership to the state. In 1986 Pacific Telesis was the first company to become involved in the telecommunications sector under the BTO scheme. It was granted a 10-year concession by CAT to provide a paging service. By

1999 the number of BTO concessions granted to private firms had reached 30 (TDRI 1999).

A particularly complicated saga concerning the granting of licences to private firms in the fixed-line sector occurred not long before the BTO system was introduced and was a factor leading to its creation. In 1984 the government under Prime Minister Prem Tinsulanonda allowed a group of Canadian investors to develop a digital network in Thailand. However, within three years this permission was withdrawn by the same government on the grounds that it violated Thai law. Under the TOT Act non-government entities are permitted to operate communications services only for their own use. It was therefore deemed that if the Canadian investors insisted on continuing to install the network, they would be the only ones allowed to use it (Soonthonsiripong 1998).

Severe supply shortages during the economic boom of the late 1980s and early 1990s created further opportunities for private firms to participate in the fixed-line sector. In 1990 the TOT sought to address the shortage problem by granting a concession for the 'Three Million Telephone Lines Expansion Project'. Under the original plan, the private firm that won the bid for this project was to be allowed to operate, for 25 years, 2 million lines in the BMA and 1 million lines in the PTAs. The company that won the bid in 1990 was TelecomAsia (TA). However, in 1991 the Thai government, headed by Prime Minister Anand Panyarachun, cancelled the project in its then current form on the grounds that it created a monopoly for TA (Soonthonsiripong 1998). The government recast the Three Million Lines Project by dividing it into two separate projects. Consequently, TA and TT&T were granted licences for these two projects as discussed in sub-section 2.1 above.

The examples of the Thai government reneging on telecommunications contracts with the private sector, as described in the two previous paragraphs, are, unfortunately, not isolated cases. They serve to underline the urgent need for legislative reform and the appointment of an independent regulator for the industry. These issues are taken up in section 4.

Apart from granting concessions, the government has allowed foreign investors to participate in the provision of communications in Thailand, but with three overriding restrictions. First, foreign investors are not allowed to establish their own communications network, so to provide their services they are required to lease circuits from the telecommunications SOEs. Second, the shareholding by foreign investors in a telecommunications project must not exceed 40 per cent of the total. Third, foreigners working on telecommunications projects must bring professional skills to those projects.

Increasing the number of telecommunications operators through the BTO system has helped overcome the problem of supply shortage. However, even though there are now more operators in the field than before, the BTO system has not produced a truly competitive telecommunications market. This is because TOT and CAT as regulators retain powers governing price setting. Thus it can be said that the BTO regime has produced only the 'illusion' of a competitive market. In practice, the state still exercises extensive control, which denies the potential for price competition in the market.

3.2 Long-Term Plan

Economic planners in Thailand have learned from the experience of developed nations that competitive markets can produce a number of benefits, including cheaper prices, improved service and a supply of goods and services sufficient to meet demand. Largely for this reason in 1997 the Thai government signed an agreement with the World Trade Organization (WTO) to liberalise fully the telecommunications market by 2006. Following liberalisation the market structure of Thailand's telecommunications will have three tiers: an inter-carrier or primary market, a resale market and a retail market.

The WTO agreement that Thailand and other nations have signed requires the signatories to commit themselves to a certain date for full liberalisation of the telecommunications sector. Those countries that are not able to liberalise immediately must commit themselves to a schedule of progressive reduction of restriction to competition. In recent years some sectors of the Thai telecommunications industry, such as consulting services and equipment sales and service, have been opened for free market entry. Other sectors such as database access services and teleconferencing are still restricted. The precedents set by other nations moving to open their telecommunications markets offer valuable lessons for Thailand in preparing for more advanced stages of reform.

Major aims of market liberalisation are to abolish monopoly power and increase private sector participation and investment. Steps towards these aims in Thailand's telecommunications have been undertaken. One move was to revise the Thai Constitution in 1997 to accommodate the government's agreement with the WTO (Nanakorn 1998). While the government has made a start with telecommunications liberalisation, many tasks remain to achieve full liberalisation by 2006. Two of the most important are shifting ownership of the BTO operations and reforming the telecommunications legislation.

4. TOWARDS LIBERALISATION

4.1 Telecommunications Concession Conversion

International experience suggests that establishing a clear market structure before or early in the process of liberalisation increases investor confidence and so facilitates reform. One of the initial tasks for the Thai government is therefore to restructure the current BTO concession arrangements.

The committee on converting BTO concession agreements was formed early in 2000 by the State Enterprises Policy Committee. The Thailand Development and Research Institute (TDRI) was appointed to act as consultant to the committee. Although, as mentioned above, 30 concessions under BTO agreements were granted from 1989, by 2000 less than 20 remained current. The number decreased with the economic crisis in 1997 causing some of the BTO concessionaires such as Radio Phone, Fonepoint and Lines Technology to close down. Shinawatra Datacom and Shinawatra Directories had already been converted into joint ventures. In September 1997 TOT formed a joint venture with Shinawatra Datacom in which it became a 23.5 per cent shareholder, and joined Shinawatra Directories as a 49 per cent shareholder in July 1998. At the time of the conversion of these concessions the major equity holders in both Shinawatra Datacom and Shinawatra Directories were interests associated with Thaksin Shinawatra, who was elected Prime Minister of Thailand in February 2001.

The first phase of the conversion project followed a conceptual framework drawn up by TDRI. The most difficult task for the conversion committee to date has been determining the amount of compensation to be paid to the concessionaire companies. This determination is contentious because a number of variables are considered relevant in the calculation, such as future growth, consumer behaviour under a competitive environment, and the diffusion of new technology. Under the TDRI conversion framework, the length of the concession period has become a vitally important factor since the amount of compensation is to be calculated from the date of the conversion until the end of the concession period (TDRI 1999).

At the beginning of 2001, after prolonged discussions between the three main bodies – the concession conversion committee, the concessionaires and the concessionees – only a few concessions were considered to be potentially viable. These were between:

- TOT and TT&T (a provincial fixed-line carrier);
- TOT and TA (a metropolitan fixed-line carrier);
- TOT and AIS (one of the largest mobile carriers);
- CAT and TAC (a cellular mobile carrier); and
- CAT and CP Orange (a new cellular mobile operator).

A number of key principles recommended by TDRI were applied in the conversion process. These principles were important shapers of the emerging free market in Thailand's telecommunications (TDRI 1999):

- *Principle 1*: The market for each telecommunications service is to be free from the monopoly power of either a sole operator or a single dominant operator.
- *Principle 2*: To ensure fair competition, all players (both existing operators and new entrants) are to be subject to a level playing field (the same rules and regulatory restrictions), and are to have similar rights and obligations.
- *Principle 3*: To protect against collusion, no operator is to have a vested interest in any other Thai telecommunications operation, either by way of cross-shareholding or revenue-sharing arrangement.
- *Principle 4*: Each private concessionaire is to receive benefits at the same level as those that can be expected from the specified terms and conditions in the concession agreement for the whole concession period.
- *Principle 5*: The value of compensation is to be calculated under the following guidelines: (a) private concessionaires will be compensated if their revenue streams decrease as a result of concession conversion, and, conversely, concessionaires will compensate the state if their revenue streams increase; (b) the specific details of each individual contract as a result of liberalisation will be included in the calculation of future revenue streams; and (c) the impact of normal business risk, such as economic slump, business strategy or market factors, will not be taken into account in calculating the value of compensation.
- *Principle 6*: Both parties to a contract are to agree on the terms for concession conversion.
- *Principle 7*: The private concessionaire will be awarded an operating licence covering the same type and scope of services as those specified in the BTO concession. The concessionaire will also be compensated for any additional obligations that are incurred over and above those specified in the concession. However, the concessionaire will bear responsibility for any obligation relating to any additional future rights beyond those specified in the concession contract.
- *Principle 8*: The Thai government will determine and implement measures to decrease limitations upon and obstacles to effective competition between TOT and CAT in markets where they are already competing either directly or through their private concessionaires.

Unfortunately the conversion process stalled. Not one concession agreement was converted by the end of 2000 when the term of the conversion committee was originally scheduled to expire. Even though its term was extended for two months, still not one agreement was completed. There was a fundamental problem in that the parties to the concession contracts could not agree to the terms of conversion established by TDRI. Principles 4 and 5 relating to compensation to be paid to the concessionaire companies were the main points of contention among the parties and the major obstacles to the conversion process. Nevertheless, an encouraging sign has emerged from the new government regarding this matter. Soon after its election it commissioned the Intellectual Property Institute at the Chulalongkorn University to develop a revised set of principles aimed at achieving agreement among the parties regarding the terms for compensation (Telecom Journal 2001b), and so finally bring about a resolution of the conversion process.

4.2 Legislative Reform

In addition to converting the BTO concessions, legislative reform is also crucial to the liberalisation process. It is commonly accepted that whatever its appearance, the Thai telecommunications market is not truly competitive. The telecommunications legislation has actually impeded the development of a competitive market. In particular, the CAT and TOT acts serve as crucial barriers to the entry of new firms into the industry (Soonthonsiripong 1998). Consequently, changing this legislation so that it allows free entry into the market has been identified as one of the major tasks in the Master Plan for State Enterprise Sector Reform ('Master Plan') (MOF 1998).

4.3 Reform of State-owned Enterprises

It is generally recognised that the government's role in the market through its SOEs is one of the major reasons for the relatively retarded development of telecommunications in Thailand. The Thai parliament thus debated at length the Corporations Act that was approved in 1999, and which will serve as the foundation for corporatising the nation's SOEs. The Corporations Act sets out a three-step process to privatise all SOEs. First, they are to be transformed into limited public companies. Next, with the assistance of international consulting firms, they will select strategic partners, that are likely to be foreign investors. The former SOEs and the consultants will then draw up implementation plans and new management strategies. Finally, the corporatised entities will be privatised.

The major goal of the reform programme is to improve the efficiency of the business operations conducted by SOEs so that they can perform successfully in a competitive market environment. To pursue this goal the State Enterprise Reform Commission (SERC) was established in 1999 by merging the Corporatisation Committee and the State Enterprise Privatisation Commission. The major responsibility of the SERC is to review and propose amendments to plans to privatise SOEs, drawn up by the SOEs themselves within terms established by SERC. In this process proposals and suggestions are to be invited from the private sector, and the privatisation plans are then to be submitted by SERC to Cabinet for final approval.

By mid-2001 the government had revised the SOE corporate governance system. Before implementing the next steps of corporatisation and privatisation, the Good Enterprise System and the Performance Agreement System were adopted to evaluate the performance of SOEs. However, these two systems were recognised as ineffective and a method was being considered to integrate these two systems into a 'balanced scorecard'.

4.3.1 Corporatisation

Concrete discussion of the post-corporatisation structure of SOEs, particularly the TOT and CAT, began in 1997. The dominant proposal was for the formation of two companies, a 'TOT company' and a 'CAT company', as subsidiaries of a holding company, initially to be fully owned by the Ministry of Finance (MOF). This proposal was finalised in 1999 under the government of Prime Minister Chuan Leekpai. The Corporatisation Act to legislate this proposal was approved in July 1999 (Soonthonsiripong 2000). Although under the initial schedule in the Master Plan it was anticipated to have this structure in place by October 2000, the expected completion date for this stage was postponed to 2001, and then again to 2002. The reason for these delays was that identifying a clear organisational structure for the holding company and the roles and responsibilities of the subsidiary companies, had not been finalised.

4.3.2 Privatisation

Efforts to privatise Thailand's SOEs have been under way since the early 1960s. The number of SOEs was reduced from more than 100 in 1961 to 59 in 1998 (MOF 1998). Extant SOEs by mid-2001 can be categorised into five major sectors: telecommunications, water, energy, transport and 'others'. The dominance of SOEs in each sector enables them to enjoy monopoly or quasi-monopoly powers. Conceptually the privatisation programme is designed to strengthen the Thai economy and to provide the basis for the nation's continued economic growth. A well-designed privatisation programme is seen as the best way to restructure the SOEs and increase capital investment.

Privatisation of SOEs is to be conducted under the Corporatisation Act 1999. Following corporatisation the TOT and CAT companies will invite private firms to engage as strategic partners. With the participation of strategic partners, MOF ownership of each of the TOT and CAT companies will be reduced to 49 per cent. A maximum of 25 per cent of the TOT and CAT companies' shares will be available to foreign investors. For TOT, 4 per cent of shares will be sold to TOT employees, and 22 per cent will be offered to private Thai investors, none of which will be allowed to hold more than 5 per cent of total equity. A consortium comprising Arthur Andersen Consulting and Deutsche Bank has been chosen to advise TOT on this process (Ackerholt and Hartford 2000).

The structure and ownership of the privatised CAT company will differ somewhat from TOT. The telecommunications department of the CAT company will become a limited public company to be called CAT Telecommunications Company Limited, while its postal department will form a separate entity, CAT Postal Company Limited. The MOF will initially own all shares in both of these entities. As with TOT these two firms will invite private strategic partners to take up a maximum 25 per cent of shares. Domestic investors will be offered around 22 per cent in each of the two companies, and a minimum of 3 per cent will be reserved for staff.

The Corporatisation Act provides that the final step of privatising telecommunications in Thailand will be to reduce government ownership in the public companies to less than 30 per cent, and to list the shares available to the public on the Thai stock exchange. This process has been scheduled in the Master Plan to take place between 2004 and 2005 (Ackerholt and Hartford 2000).

4.4 An Independent Regulator

A primary objective of independent regulatory bodies is to facilitate the operation of market competition. For telecommunications this means that policies – such as for interconnection, numbering and universal service – will be introduced to guarantee free and fair competition, and to ensure high standards of service. The regulator also has the responsibility to issue licences to operate telecommunications facilities on non-discriminatory terms (Ypsilanti and Xavier 1998).

The roles and responsibilities of policy-makers, regulators and operators in Thailand currently overlap among different government bodies. The streamlining of responsibility through an independent regulator is therefore crucial to the liberalisation process. Under the objectives stated in the Master Plan, a National Telecommunications Commission (NTC) is to be established as an independent body to ensure the existence of free and fair competition among existing operators and new players. The government under former Prime Minister Chuan Leekpai acknowledged that an independent regulator with clear objectives and responsibilities was needed to confirm the transparency of the liberalisation process and of regulatory decisions after liberalisation is achieved.

In mid-2001 moves were under way to establish the NTC as Thailand's independent telecommunications regulatory body. This recognises that the market mechanism alone will not maximise the interests of consumers or guarantee effective competition. The NTC is to have four broad functions: drafting the terms and conditions of licences; regulating the price mechanism (for example,, price cap regulation); defining rules to ensure fair and transparent competition; and providing policy advice to government. Cabinet will continue to play the major role in policy-making through setting long-term goals for the sector, defining social obligations and setting the broad direction of pricing policy (MOF 1998).

It has been determined that membership of the NTC is to include specialists selected from both the public and private sectors. The Thai Senate is to select a membership of seven for appointment to the NTC from a list of 14 nominees selected by the Prime Minister. NTC staff members are to include members of the professions, for example, lawyers, economists, financial analysts and accountants. Funding to support the regulatory body is to be derived primarily from a levy on operators (MOF 1998).

5. CONCLUDING COMMENTS

It is possible to identify some positive results from the initial moves Thailand has taken towards increasing competition in its telecommunications industry. The participation of private firms in the market for basic services has produced some degree of competition, which has given rise to price reductions and other benefits to consumers. Significant reductions in installation charges, a single rate for domestic long-distance calls, and free internet usage have been offered by some firms as incentives to increase fixed-line subscriptions. However, because call prices are set by TOT and CAT, there is as yet no effective price competition in the industry.

In the cellular market, free entry for the sale of handsets has generated significant price reductions since 1999. Nevertheless, handsets are still more expensive in Thailand than in developed countries. In 2001 there were indications that the degree of competition in the mobile market was likely to increase, which should result in lower call charges.

Competition across sectors has been found to emerge between the cellular and fixed-line markets. The personal communications telephone (PCT) service was launched late in 2000 by Asia Wireless Communications, a company related to TA that operates the 2.6 million fixed lines in the BMA. Although the PCT was introduced as a value-added service of fixed-line networks, because of its mobility function it has provided competition for regular mobile services. Price reductions for PCT handsets have enticed a number of cellular subscribers to switch to fixed lines. Some mobile phone operators have reacted promptly to PCT developments by offering pricing packages aimed at maintaining and increasing mobile usage.

This chapter has shown that from the late 1990s the Thai government has taken steps towards the reform of the nation's telecommunications market. More change is to come as the privatisation process moves to its scheduled completion in 2006, in line with Thailand's agreement with the WTO. The industry has moved from complete state regulation, with ownership by a number of SOEs, to a more open market system, especially as new products have entered the market.

Among the challenges still facing the industry is the need to increase significantly the nation's teledensity rate, and to better balance supply of new lines with demand in both metropolitan and provincial areas. The entry into the industry of private corporations with finance and expertise is working to deliver better services and prices for consumers and a more vibrant market. Thai policy planners continue to learn from the precedents of developed nations and from their own experiences of corporatisation and privatisation.

Reform of Thai telecommunications has been subject to a number of obstacles and delays in recent years, most notably in relation to resolving the concession conversion process and establishing an independent regulator. There are encouraging indications, however, that the new government is determined to clear the gridlock and set the telecommunications industry back on the road to liberalisation.

ABBREVIATIONS

AIS	Advanced Info Service Plc
BMA	Bangkok Metropolitan Area
BTO	build–transfer–operate scheme
CAT	Communications Authority of Thailand
CAT Act	Communications Authority of Thailand Act 1976
GSM	global system for mobile communications
MOF	Ministry of Finance
MOTC	Ministry of Transport and Communications
NTC	National Telecommunications Commission
PCN	personal communications network
PCT	personal communications telephone
PTAs	provincial telecommunications areas
PTD	Post and Telegraph Department
SERC	State Enterprise Reform Commission
SOEs	state-owned enterprises
TA	TelecomAsia

TAC	Total Access Communications
TDRI	Thailand Development and Research Institute
TOT	Telephone Organisation of Thailand
TOT Act	Telephone Organisation of Thailand Act 1954
TT&T	Thai Telephone and Telecommunications
VoIP	voice over internet protocol
WTO	World Trade Organization

REFERENCES

Ackerholt, P. and H. Hartford (2000), *Telecommunications in Thailand: A Study of the Ongoing Liberalization Process*, Linköping University: International Business Programme.

MOF (1998), *Master Plan for State Enterprise Sector Reform*, Bangkok: Royal Thai Government.

Nanakorn, P. (1998), *Current Plans for the Reform of the Laws Relating to Telecommunications: Preparation for Fair and Free Competition in Distribution of Frequencies Pursuant to Section 40 of the Constitution of the Kingdom of Thailand,* Bangkok: TDRI.

Soonthonsiripong, N. (1998), 'Regulatory reform of telecommunications in developing countries: A case study of the fixed-line telephone network in Thailand', PhD thesis, University of Adelaide: School of Economics.

Soonthonsiripong, N. (2000), 'Telecommunications reform in the Asia Pacific: Can the developing countries benefit from the experience of the developed countries?: A case study of Thailand', Telecom Workshop, Jakarta, 21–22 November.

TDRI (1997), *The Telecommunications Development Plan 1997–2006*, Paper proposed to the Ministry of Transport and Communications, Bangkok, TDRI.

TDRI (1999), *The Preliminary Report of the Telecommunications Concession Conversion Study*, June, Bangkok: TDRI.

Telecom Journal (2001a), 'TOT and CAT to decide on the format of 1900 MHz project', no. 367, 30 July–5 August, Bangkok.

Telecom Journal (2001b), 'TDRI framework for concession conversion to be revised for next round of negotiation', no. 367, 30 July–5 August, Bangkok.

TOT (1991), *Annual Report 1991*, Bangkok: Division of Statistics and Report.

TOT (2001), *Status of Telephone Services: June 2001*, (http://www.tot.or.th).

Yoonaidharma, S. (1993), 'Role of laws over the Communications Authority of Thailand operation', *Future Directions of the Communications Authority of Thailand*, Appendix A, Bangkok, TDRI.

Ypsilanti, D. and P. Xavier (1998), 'Towards next generation regulation', *Telecommunications Policy*, **22** (8), 643–59.

8. Sri Lanka: From telecommunications privatisation to anti-competition?

Sisira Jayasuriya and Malathy Knight-John

The fundamental and rapid changes being experienced by telecommunications markets throughout the world, brought about both by technological developments and by competition-oriented reform policies, also are reflected in the changes made to Sri Lanka's telecommunications industry since the 1980s. These changes have been driven by some major government initiatives to open up the sector to new entrants and greater competition. With the restructuring of state-owned telecommunications, the deregulation of value-added telecommunications services, and the entry of two new players into the fixed-access sector in 1996, Sri Lanka's telecommunications industry has been profoundly transformed, and has become one of the most open and competitive among developing countries in the Asia–Pacific region.

But since the partial privatisation of the restructured state-owned entity in 1997, there is evidence of significant resistance from within the government itself to pro-competitive reforms. This chapter examines and presents the state of Sri Lanka's telecommunications industry, the regulatory policy framework, and the pertinent competition policy issues in the sector, with particular attention to the government's apparent shift away from its earlier strong pro-competitive stance.

1. HISTORY AND RATIONALE FOR TELECOMMUNICATIONS REFORM

The changes in the telecommunications sector in Sri Lanka are directly linked to the recognition that an improved telecommunications sector is essential to achieve faster economic growth, and also to a policy shift towards a greater role for the private sector in all areas of the economy, including infrastructure services.[1] Telecommunications was a very backward sector of the economy right up to the 1990s. This was not accidental. Until quite recently, infrastructure development in Sri Lanka was largely the preserve of the state – a historical legacy from the British colonial era. However, under-investment in infrastructure, other than in irrigation and, to some extent in hydro-power, was a notable feature of post-independence Sri Lanka, with short-term political benefits, rather than long-term development imperatives guiding infrastructure investments. Even after radical policy reforms were initiated in 1977 to shift the economy in a liberal, market-driven and export-oriented direction, large-scale infrastructure investment was almost entirely concentrated on a few selected projects and did not extend to the telecommunications sector.[2]

The deficiencies of the telecommunications sector became increasingly apparent with an increase in business activity in the post-1977 years. There was high but unmet demand for telecommunication facilities, with long waiting periods, and poor transmission quality and system reliability (Jayasuriya and Knight-John 1998). Clearly there was inadequate investment in the sector, but not because the telecommunications sector itself was unprofitable. The shortfall in investment instead was the product of overall government fiscal constraints and investment priorities.[3] The impetus for telecommunications reform, as in other infrastructure industries, arose from recognition that the benefits of policy liberalisation in terms of export growth and attraction of foreign investment could only be realised if adequate infrastructure services were available. An efficient and modern telecommunications network was obviously essential if Sri Lanka's potential as a regional financial, commercial and communications hub was to be achieved. Added pressure for improved telephone and fax services came from rural areas with the very large numbers of rural workers now employed overseas.

But the state, saddled with large fiscal deficits, could not undertake the required investments with its own resources. Nor was it able to tap foreign aid or foreign borrowings as in the past. The ideological shift in the donor community had moved towards private sector funding of infrastructure investments in sectors such as telecommunications; the donors also disapproved of government borrowings from international capital markets for such investments. Their preferred strategy for infrastructure development was via BOO/BOT type arrangements and strategic public–private partnerships (Knight-John 1997). The government's own policy orientation favoured private sector initiatives in potentially profitable services sectors whenever that was politically feasible. This combination of factors led to a range of institutional and policy changes in the telecommunications sector from the early 1980s onwards.

2. OWNERSHIP AND INSTITUTIONAL CHANGES IN THE TELECOMMUNICATIONS INDUSTRY

Reforms in the telecommunications sector have had three interrelated strands: the restructuring of the state-owned telecommunications entity, the deregulation of value-added services, and the establishment of regulation.

2.1 Restructuring of the State-Owned Telecommunications Entity

Until 1980, the telecommunications industry was organised on the lines of the Post, Telegraph and Telephone (PTT) model, with the state owning and operating both telecommunications and postal services. The first major reform involved a de-linking of the two services. This was a decision consistent with international experience that rapid technological improvements in telecommunications tended to make the industry functionally incompatible with the less dynamic postal sector.

But even after the de-linking of the two sectors, telecommunications continued to be inefficient. The state-owned industry functioned as a monopoly shielded from competitive pressures, and gained no significant injection of funds. Although limited steps toward competition were introduced in some elements in 1981 (paging) and 1985 (customer premises equipment), these were of marginal importance to the operations of the state enterprise. Moreover, attempts to restructure the incumbent Department of Telecommunications (DOT) failed because of the climate of political instability during the late 1980s and trade union opposition to the restructuring of state enterprises. The conversion of the DOT into a government corporation, Sri Lanka Telecom (SLT), in 1991 did not change its basic administrative and financial dependence on the government, since the radical alternative of privatisation was not seriously raised in policy discussions at this time. Indeed, the international paradigm shift towards privatisation of public utilities had not yet developed the momentum it was to acquire later, and the political obstacles to privatisation in Sri Lanka were severe.

Pervasive government interference in the operations of SLT prevented it from functioning as a commercial enterprise, and it was constrained for investment funds because government priorities lay elsewhere. In 1993, the government set up a state-owned company, Sri Lanka Telecom Services Ltd (SLTS) as a subsidiary of SLT with more financial and administrative independence to accelerate network rollout. However, there was little improvement in service quality and both corporatisation and the creation of a subsidiary entity were seen as threats by elements within SLT (Samarajiva 2000b).

It was widely expected that radical restructuring of state enterprises would cease with the election of a new government in late 1994 that included 'socialist' and 'communist' parties. Popular opposition to many privatisation measures of the previous regime had been exploited by the opposition during the elections and the new government had promised in late 1994 that there would be no privatisation. No plans for such privatisation were included in the President's Economic Policy Statement in January 1995 (see Government of Sri Lanka 1995), but the new government's actual economic policy agenda was quite different to its public proclamations during the elections. In fact it proceeded to accelerate the policy of liberalisation process, particularly in the area of privatisation, and soon went public with restructuring plans for telecommunications that were clearly aimed at privatisation – though the term privatisation was initially avoided because of political sensitivities. It exploited its close political links with trade union leaders to effectively neutralise opposition to privatisation and was helped by the fact that, by this time, the privatisation of public utilities had become ideologically accepted across the country's broad mainstream political spectrum, including the 'socialist' and 'communist' parties in the government.

In 1996, SLT was converted to a public company, Sri Lanka Telecom Ltd (SLTL), as an initial step toward partial privatisation. In 1997, the government sold 35 per cent of its shareholding in SLTL for US$225 million to Nippon Telegraph and Telephone (NTT), the incumbent state-controlled operator in Japan (with a reputation for anti-competitive behaviour according to the *Economist*, 22–28 July 2000). The government retained 61.5 per cent share ownership and SLTL employees were given the remaining balance, 3.5 per cent. The employee share ownership plan was used to counter anticipated union resistance to privatisation. However, some employees later sold their shares to NTT so that in late 2000 it held 35.2 per cent of SLTL shares.

The government also signed a five-year management agreement with NTT to provide technical and management expertise to SLTL. The agreement signed with NTT, along with subsequent changes to the licence of the incumbent operator, committed the government to issuing no further licences for wireline or international telephone services until August 2002, and also the creation of a tariff re-balancing programme (Samarajiva 2000b). The agreements also suggested that SLTL would enjoy a monopoly in international telephony until 2002.

SLTL's performance in terms of network expansion and modernisation, quality of service and customer relations, and internal operational efficiency improved with privatisation. For instance, new connections increased from 72 457 in 1997 to 143 075 in 1998; the average waiting time fell from seven years to less than one year; call completion rates rose to around 40 per cent; and fault clearance improved to 52 per cent (Duff & Phelps Credit Rating Lanka Ltd 2000). This performance, together with optimism about future growth and profit prospects, raised SLTL's profile among investors. A US$1.5 million debenture issued by SLTL in March 2000 to finance its expansion programme was rated AA+ by Duff and Phelps Credit Rating Lanka Ltd and over-subscribed on its opening day. Indeed, this debenture issue was hailed as a landmark, being the single largest rated and listed debenture by an unlisted company in Sri Lanka's history.

This encouraged the government to move to undertake further sales of its shares to bring its stake below 51 per cent, with part of the sale to be listed in the Colombo market and the remainder in a foreign market. It contemplated a private placement of 10.5 per cent of its shareholdings, as a preparatory step, prior to the Initial Public Offering (IPO) with the aim of bringing in technology-driven partners who would add value to the company. Extensive negotiations towards this end were held with Singapore Technologies, a subsidiary of Singapore Telecom, and the second largest company in Singapore (next to Singapore Telecom). However, this was unsuccessful because Singapore Technologies wanted a bigger management role than either NTT or the government were willing to concede. The IPO was re-scheduled for June/July 2000 and was expected to raise around US$400 million. But this has now been indefinitely postponed, with the government blaming increased country risk due to the intensification of the civil war and the plunge in international technology and telecommunications stocks for the postponement.

3. COMPETITIVE REFORMS AND THE DEVELOPMENT OF THE REGULATORY FRAMEWORK

The major changes in the industry came initially from pro-competitive reforms that progressively deregulated value-added services. These started in 1981, with the entry of private paging operators who started services initially without formal licences and were only later brought into the regulatory system.

In the first phase of deregulation, from 1981 to 1995, competition was introduced only on the periphery of the industry. Despite this cautious approach, around 20 system operators – including the incumbent operator – were licensed during this period, but only after the enactment of the Sri Lanka Telecommunications Act No. 25 of 1991. The second phase of competition

Table 8.1 Licensed telecommunications system operators, as at 11 May 2000

Service	Operator
Fixed telephony	Sri Lanka Telecom Ltd
	Suntel (Pvt.) Ltd
	Lanka Bell (Pvt.) Ltd
Mobile telephony	Lanka Cellular Services (Pvt.) Ltd
	Mobitel (Pvt.) Ltd
	MTN Networks (Pvt.) Ltd (Dialog GSM)
	Celltel Lanka Ltd
Facilities-based data communications services[a]	Lanka Communication Services (Pvt.) Ltd
	Electroteks (Pvt.) Ltd
	SITA-Société Internationale de Télécommunications
Aéronautiques	Lanka Internet Services Ltd
	Ceycom Global Communications Ltd
	ITMIN Ltd
Non-facilities-based data communications services[b]	Millanium Communications (Pvt.) Ltd
	MTT Network (Pvt.) Ltd
	Metropolitan Telecom Services (Pvt.) Ltd
Internet-based data services (ISPs)	Eureka Online (Pvt.) Ltd
	Pan Lanka Networking (Pvt.) Ltd
	Projects Consultants International (Pvt.) Ltd
	DPMC Electronics (Pvt.) Ltd
	Celltel Lanka Ltd
	Dynaweb Services (Pvt.) Ltd
	Victra-Soft (Pvt.) Ltd
	East West Information Systems Ltd
	Lanka Global Online (Pvt.) Ltd
	Visual Internet (Pvt.) Ltd
	Dynanet Ltd
	MTN Networks (Pvt.) Ltd
Public payphone services	The Payphone Co. (Pvt.) Ltd
	TSG Lanka Ltd
Paging services	Infocom Lanka Ltd
	Bell Communications Lanka (Pvt.) Ltd
	Fentons Ltd
	Intercity Paging Services (Pvt.) Ltd
	Equipment Traders (Pvt.) Ltd
Trunked radio	Dynacom Engineering (Pvt.) Ltd
Leased line services	MTT Network (Pvt.) Ltd
Other	Air Lanka Ltd

Notes:

[a] Facilities-based operators can set up their own transmission facilities (any line, radio, or other electro-magnetic or optical system or any other similar technical system)

[b] Non-facilities-based operators cannot establish their own transmission facilities, but can use the facilities of other licensed operators to connect subscribers and sites

Source: Telecommunications Regulatory Commission of Sri Lanka

saw the 1996 entry of two wireless loop operators (WLLs), Suntel and Lanka Bell, brought in to compete directly with SLTL in the fixed-access market. Suntel is a joint venture between Telia of Sweden, Townsend Ltd of Hong Kong, and the Metropolitan Group of Companies and the National Development Bank of Sri Lanka. Lanka Bell is a joint venture between Transmarco of Singapore, an infrastructure fund, and Nortel, an equipment supplier. These operators were given duopoly status in WLL services until the year 2000, with the possibility of a five-year extension if they met performance targets, such as the provision of 100 000 lines by the year 2000. But their duopoly status was eroded when SLTL also was permitted to operate WLL services within a specified range of frequencies (Samarajiva 2000b). The industry status in the sector in late 2000 is shown in Table 8.1.

3.1 Establishment of Regulation

The foundation for telecommunications regulation in Sri Lanka was laid with the enactment of the Telecommunications Act No. 25 of 1991. This legislation led to the trifurcation of the sector – operational functions went to SLT; regulation becomes the responsibility of the Office of the Director General of Telecommunications (ODGT); while telecommunications policy remained within the Ministry.

The ODGT was a single-person authority, modelled on the lines of Britain's OFTEL, with the Director General handling regulatory issues and advising the Minister on policy matters. However, the ODGT, a government department with low remuneration, was unable to attract well-qualified employees. With a shortage of skilled workers, ODGT could not carry out many of its key regulatory functions, such as consumer protection, public hearing procedures, and enforcement of licence conditions. Moreover, the separation of regulatory and operational functions, as envisaged in the 1991 Act, did not materialise in practice as both the ODGT and SLT reported to the Ministry.

A draft amendment to the 1991 Act was prepared in 1993 to address some of the problems faced by the regulatory authority. The draft created a more independent agency by reducing Ministry control over the ODGT, but the plan was not accepted by the government. However, legislation was introduced in 1996, by way of the Sri Lanka Telecommunications (Amendment) Act No. 27, that converted the ODGT to the Telecommunications Regulatory Commission (TRC). This law replaced the single-person authority with a five-member Commission comprised of three part-time members with security of tenure, and two ex-officio members. The part-time members were drawn from law, finance, and management; the two ex-officio members were the Secretary of the Ministry, serving as Chairman, and the Director General of Telecommunications, serving as the Chief Executive Officer of the Commission.

TRC was expected to follow the broad objectives set out in a 1994 government telecommunications policy document. Its responsibilities included advising the government on licensing and on tariffs, pricing and subsidy policies; determining, in consultation with the Minister, tariffs and methods for calculating tariffs; approving interconnection charges where operators reach a mutual agreement on these charges and determining the charges in the absence of such agreement; functioning as the sole manager of the frequency spectrum; ensuring that operators comply with quality standards specified in the 1996 Act; and protecting consumer interests. It also was expected to approve the types of telecommunications equipment used by operators to ensure network compatibility.

While the TRC was supposed to operate within the government policy framework, its mandate to advise the Minister also allowed it to exert some influence on government policy. The objectives of the 1994 policy document included provision of telecommunications facilities to all at cost-based tariffs, achievement of universal service provision of an acceptable quality of service, elimination of waiting lists, and protection of defence, security and environmental interests of the country. Unlike its predecessor, the TRC did not have to depend on state funds but could draw on licence and other fees to finance its operations. However, the independence of the regulatory authority was limited by the appointment of the Secretary to the Ministry as the ex-officio Chairman of the Commission. Its independence was further compromised when it absorbed several former employees of the SLT.

A new and more independent phase of regulation came in 1998 with the hiring and training of new personnel and the initiation of several pro-competitive regulatory measures. The TRC's stance signalled that competition, rather than excessive regulation, should be the driving force in the industry during this period. TRC's mission statement, developed at this time, also emphasised consumer interests and competition (Samarajiva 2000a). The government decision to commit to the Regulatory Reference Paper, a part of the Fourth Protocol to the General Agreement on Trade in Services of the World Trade Organization (WTO), signalled a similar commitment. Sri Lanka is the only South Asian country to have made this commitment so far. But subsequent developments, to be discussed later, demonstrate that the goal of implementing pro-competitive policies has not yet materialised.

3.2 Structure and Performance of the Industry

The telecommunications sector, is, for the purpose of regulatory policy, differentiated according to market structure and technology, a formal monopoly in the fixed wireline sector (SLTL), a duopoly in the fixed wireless sector (Suntel and Lanka Bell) and competition in the mobile phone sector (four mobile operators). But many of these services are close substitutes and often complementary services. Furthermore, service providers often operate in more than one segment of the industry. Indeed, the development of convergence-friendly technologies enhances incentives for this to occur. Hence, the actual industry structure is more complex, and market segments cannot be treated as independent for analytical or regulatory purposes.

SLTL remains the dominant player with extensive market power in the overall telecommunications market, controlling approximately 60 per cent of the total industry.[4] Although its main focus of operations is the fixed-access market, it is also the leading internet service provider (ISP), operates its own payphones, and provides leased circuit and data communications services. SLTL also controls 40 per cent of one of the mobile phone firms, though the 1994 national policy on telecommunications specifically states that fixed-access operators will not be permitted to provide mobile services. By Sri Lankan standards, SLTL is a massive corporate entity; its revenue in 1999 was equivalent to 1.7 per cent of the country's GDP (Central Bank of Sri Lanka 1999).

In the fixed-access market, where it has a formal monopoly in wireline services, SLTL has a 86.3 per cent market share, whilst Suntel and Lanka Bell have market shares of approximately 9.59 and 4.11, respectively.[5] According to government sources (*The Daily News* (Colombo) 23 February 2000), the three fixed-access operators have invested around US$600 million in the past five years and more investments have been foreshadowed. Having initially drawn most subscribers from residential customers, they are now attempting to make inroads into the corporate and business segments by offering bulk discount packages and a range of sophisticated services, sometimes in conjunction with foreign firms. For example, Suntel has recently targeted corporates and small and medium businesses by linking up with Airspan Networks, London, and providing high-speed access to the internet and to public and private data networks.

The performance in the fixed-access market has been mixed. Based on Central Bank of Sri Lanka and TRC statistics, we estimate that fixed phone teledensity increased from 0.73 in 1991 to 1.8 in 1997, and then to 3.5 in 1999. Approximately 55 per cent of these fixed-access lines are located in the Greater Colombo area, indicating a continuing urban bias in this segment (Gunawardene 1999b). Service quality has improved, and average waiting time for an SLTL connection has fallen from seven years to less than a year and WLL operators, however, have no significant waiting lists (a 1999 regulatory decision to compensate consumers for delayed connections probably helped cut waiting times; see Gunawardene 2000). The WLLs have superior performance in terms of call completion rates and faults statistics. However, they complained that they have been unable to consistently achieve the 50 per cent and above call completion rate specified in their licences, allegedly because of call blocking by SLTL.[6] The SLTL, has revamped its billing procedure, particularly after a 1998 public hearing on billing issues initiated by the TRC,[7] and introduced itemised billing in 2000. Service quality is expected to further improve with the publishing of TRC's quality of service rules, and its linking of tariff increases to service quality.

There have been more extensive pro-competitive reforms in the mobile phone market, which is characterised by greater competition than the fixed-access market. For many customers, mobiles have been a substitute for fixed lines and the sector has grown very rapidly. Its compound annual growth rate during 1995–98 was 48.6 per cent as against 36.5 per cent in the fixed-access market (ITU 1999a). Overall mobile penetration increased from 0.01 in 1991 to 1.35 in 1999, and the combined fixed–mobile teledensity reached 4.9 in 1999 (Samarajiva 2000b). Sri Lanka now has a mobiles-to-fixed ratio of around 40 : 100.

Investment in the mobile phone market over the past five years amounts to around US$420 million.[8] Intense competition amongst the mobile operators has led to a movement away from analogue to higher capacity digital networks, a proliferation of value-added products such as Wireless Access Protocol (WAP) and other internet-mobile services, extensive advertising campaigns, while prices have declined at an annual average real rate of 18 to 20 per cent. According to the International Telecommunications Union (ITU) (1999b), Sri Lanka's cost of US$17.80 for a 100-minute basket of mobile tariffs compared favourably with the average costs of US$39.69 and US$38.15 in lower-middle income countries and the world. Although mobile rates in Sri Lanka are still above fixed-access rates, introduction of home-zone pricing for mobile calls and the tariff re-balancing process in the fixed-access market is eroding this gap (Samarajiva 2000b). Furthermore, mobiles can be immediately obtained, and mobile operators offer services such as pre-paid card schemes, tri-lingual customer services, and easy international direct dialling (IDD) facilities in direct competition with SLTL (Duff & Phelps Credit Rating Lanka Ltd 2000). Nevertheless, despite the fact that mobile phone technology has a distinct advantage over basic telephony in rural areas given the difficulty in laying cables in rough terrain, the mobile penetration rate in the rural sector is low at only around 0.1 per cent.

Though there are no formal entry restrictions in other sub-sectors, such as payphones, data communications services, and pagers, these sectors are growing at a slower rate. The pagers appear to be priced too high with respect to mobile phones and the number of pagers has contracted in recent years. The dominant position of SLTL is a major issue in these markets and suspicions exist that SLTL may be cross-subsidising its payphones. SLTL also had conflicts with the large data communication service providers over the provision of enhanced voice services. Several small private telecommunications bureaux (resellers) also provide informal competition and a useful public service, though the need for regulation has arisen because of alleged overcharging and other malpractices. The TRC recently implemented a scheme to regulate the activities of these bureaux by issuing permits to the three fixed-access operators and holding them responsible for issuing certificates to these resellers.

Internet access in Sri Lanka is seen as a future growth area, but internet connectivity is still low with only 0.29 hosts per 10 000 inhabitants in 1998. It is constrained by the high cost of leased lines and computer equipment and also the low level of computer literacy.

4. REGULATION AND COMPETITION POLICY ISSUES

The transformation of the telecommunications industry from a monolithic monopoly structure to a more competitive industry has resulted in better operational efficiency and a greater degree of consumer responsiveness. But much remains to be achieved. In this section, we discuss some of the major regulatory issues in the industry, and assess the success of the policy responses to these problems.

4.1 Barriers to Entry and Exit and Anti-Competitive Strategic Behaviour

There are several barriers to entry operating in the Sri Lanka telecommunications market. These include several legal/regulatory entry barriers and others that also deter new entrants and impede competition. Some issues pertaining to actual or potential strategic behaviour on the part of the dominant operator are discussed in more detail later in the sections on specific regulatory issues.

The existing legal monopoly status enjoyed by SLTL in fixed wireline phones, the (disputed) restrictions on international calls, the duopoly in the fixed wireless sector, and limitations on licences in other sectors are the major legal or regulatory barriers to entry. The number of licences is now an immediate issue in the mobile telephony market as current restrictions on issuing licences for mobile telephony was due to expire by the end 2000. Incumbent operators claim that the mobile market is saturated and cannot sustain more operators, but international experience suggests that competition may be a better option given that there is no obvious way to determine how many mobile operators should be allowed. The dominant position of SLTL also confers on it, advantages in the other markets it operates in and may act as a barrier to greater competition.

From the consumer side, the significant switching costs that have to be incurred at present by consumers moving from one operator to another create a serious barrier to competition as they deter entry by competitors in both the fixed-access and the mobile markets. The high non-refundable connection charges in the fixed market effectively tie consumers to a particular provider's network, as does the lack of number portability.[9] Moreover, the standard problem, where the technology used by one operator is not compatible with the technology used by another, also creates a lock-in effect in both markets but may soon become less of a problem if convergence becomes a reality.

4.2 Interconnection

The importance of all-to-all (any-to-any) connectivity requires the joint use of network facilities, but also helps incumbent or dominant operators gain a strategic-competitive advantage over their rivals. In common with other countries, Sri Lanka also faces the challenge of how to establish a fair interconnection regime that ensures efficient provision of access to networks. Sri Lanka is a signatory to the WTO Regulatory Reference Paper that commits it to ensuring that interconnections with major suppliers will be provided on non-discriminatory terms, in a timely fashion, and at cost-oriented rates that are transparent and sufficiently unbundled. But the implementation has proved to be difficult.

The regulatory authority intervenes in interconnection issues only if operators cannot come to a negotiated agreement on access. In November 1996, following the failure of the SLTL and the new WLL operators to reach a negotiated settlement, the TRC issued a determination for two years on interconnection in the fixed-access market that followed the principles of sender-keeps-all for domestic calls; a 35 per cent rebate on the collection rate for outgoing international traffic from the WLL networks (but no payment for incoming international traffic); and the costs of physical links being fully borne by the WLLs (Gunawardene 1999a). Interconnection charges between the WLLs was on a sender-keeps-all basis with the two operators splitting interconnection costs on a 50-50 basis.

As in most developing countries, inbound traffic far exceeds outbound traffic in Sri Lanka. The 'no payment for incoming international traffic' clause in the 1996 determination, combined with SLTL's (disputed) exclusive gateway rights in the international segment, clearly disadvantaged the WLLs. Attempts to reach a negotiated settlement were again unsuccessful and the TRC was compelled to make a new determination in November 1998. The basic ingredients of the new determination are as follows: sender-keeps-all system to be replaced by a mutual compensation arrangement for local call termination, with the transition to the new arrangement being subject to conditions such as the provision of common directory services and the elimination of the surcharge on local calls made by SLTL customers to WLL customers; national call transit and termination also to be on the basis of mutual compensation; the WLL operators are to remit 80 per cent of the collection rate to SLTL for all international calls originated from WLL networks and SLTL is to pay a fee of Rs.9.50 per minute (approximately US$0.120) to the WLLs for international calls terminating in the WLL networks; the WLLs are to bear the full cost of physical interconnection links up to the interface connection unit and SLTL is to be responsible for providing the interface unit.[10]

Given the difficulties in obtaining acceptable cost data in Sri Lanka, the interconnection rates were benchmarked on European Union (EU), International Telecommunication Union (ITU) and Federal Communications Commission (FCC) rates. Though this new arrangement was somewhat more favourable to WLLs, none of the three parties was satisfied with it. But whilst the WLLs complied with the regulatory directive, SLTL continued with the pre-1998 arrangement and took the issue to court. SLTL later agreed to comply with the 1998 determination, pending the court decision.

The WLLs also have charged that SLTL was blocking calls originating from the WLL networks and succeeded in obtaining enjoining restraining orders on the SLTL until a full inquiry is made. This issue of unfair interconnection practices also may go to the WTO, and it also relates to another major legal dispute over the use of enhanced voice services by the WLLs as cheaper gateways for international calls. The distinction between voice telephony and data communications, which was sharper at the time the three data communications licences were issued, has been blurred by technological changes. At present, there are three facilities-based data communications operators in the market with licences to provide enhanced voice services, and they have provided this service to the WLLs. According to SLTL, its exclusive right over international voice telephony precludes any others from providing those services. The enhanced voice operators claim the right to do so on the basis that their licences were issued prior to the SLTL privatisation. To add to the confusion, the WLL licences clearly state that they can interconnect with any licensed operator. The courts have determined that SLTL had failed to establish a prima facie case that it had a monopoly. Meanwhile, one of the enhanced voice operators filed action against the Sri Lankan government at the International Centre for Settlement of Investment Disputes, claiming US$200 million in damages for contravening their investment protection agreement.[11]

The terms and conditions imposed by SLTL in the case of the fixed–mobile interconnection regime that prevailed until 1999 showed similar anti-competitive behaviour. For instance, mobile operators had to pay the higher national rate calling charges for calls terminating on SLTL's network, and full retail charges for international calls originating in their networks; they could only interconnect at one point and had to bear the full costs of physical interconnection; moreover, SLTL did not pay the mobile phone operators for calls terminating on their networks (Samarajiva 2000b). In contrast, the WLL–mobile and mobile–mobile interconnection arrangement is sender-keeps-all, with these operators sharing physical interconnection costs on a 50–50 basis.

Whilst a 1999 TRC determination on fixed–mobile and mobile–mobile interconnection addressed most of the anti-competitive elements mentioned above, the proposed implementation of a calling party pays (CPP) system in the mobile sector still hangs in the balance. Under the current mobile party pays (MPP) system, the mobile operators pay SLTL for calls terminated on its network but are not compensated for calls terminated by SLTL on their networks. As such, the mobile user bears the cost of termination in the form of an incoming call charge. For outgoing calls, mobile operators charge a fixed rate other than for intra-network calls. Fixed operators have some differential rates for outgoing calls.

Mobile operators argue that the MPP system has resulted in low call completion rates, because users prefer to keep their handsets switched off to avoid the incoming call charge. Operators also realised that CPP schemes are emerging as an international standard. However, CPP raises the problem of market power in termination charges on the supply-side, given that a customer cannot choose how a given call will be terminated even when they are charged different rates depending on where the call is terminated (Noll, forthcoming). The fixed-access operators have opposed CPP, arguing that it will impose a greater burden on fixed-access users, most of whom, they claim, are from the poorer segments of the population – a claim with seemingly little evidence to support it and, not surprisingly, disputed by mobile operators. The regulatory authority initiated a public hearing on the feasibility of CPP in June 1999, but has yet to make a conclusive determination.

4.3 Universal Service Obligations

Universal Service Obligations (USO) in Sri Lanka primarily involve the provision of telecommunications facilities to rural areas and to low income groups. Although twice as many connections have been provided in the past five years than in the entire previous century of telephony in Sri Lanka (Samarajiva 2000b), rural coverage remains low with SLTL and the mobile operators recording rural penetration rates of approximately 1.70 and 0.01 respectively and around 75 per cent of the WLLs' operations being concentrated in urban areas.[12] Rural demand for telecommunications services, however, is high, and was cited as the second most important unmet need in a rural survey conducted by the TRC.

The government's commitment to universal service is set out in the 1994 telecommunications policy document as well as in Sri Lanka's schedule of WTO commitments, where it is specified that USOs must be administered in a transparent, non-discriminatory, and competitively neutral manner. The government's main approach to universal service provision has been to connect remote post offices to the SLTL network using Treasury funds. However, this practice goes against the WTO principles that state that USOs have to be administered in a transparent and competitively neutral manner. According to the TRC, post office connections provide more limited access at a higher cost than payphones. The TRC implemented a scheme from 1999 to subsidise payphone facilities to rural and sub-urban areas using its own access and licence funds. The scheme is to be phased out when the set target of 100 payphones in each of the country's 25 administrative districts has been reached. In 1999, it recommended that all fixed and mobile-access operators bid to connect clusters of post offices, subject to a ceiling of US$5,400 (Samarajiva 1999). This recommendation is yet to be implemented.

There seems to be a clear conflict between the current situation and WTO commitments. The SLTL has no legally binding universal obligations, but the WLLs do have some USO-type commitments built into their licence agreements. SLTL statistics indicate that more connections are being provided in urban areas since privatisation in 1997 (Gunawardene 1999b). The controversy over interconnection has also spilled over into the area of universal service. SLTL has argued that because its profits from international calls are reduced when it has to pay the WLLs for international calls terminating on their networks, its ability to cross-subsidise and expand network roll-out has been diminished (this fee, however, amounts to less than 20 per cent of settlement revenues (Samarajiva 1999)).

The WLLs, on the other hand, are required to have at least ten working telephones in each secondary switching area by the end of 2000, with one telephone in the rural area being considered equivalent to ten telephones as an additional performance incentive. Indications are that they are unlikely to meet them, though they face an annual penalty of US$80,000 if they fail to meet these targets (a penalty too low to force compliance).

The question of who should be responsible for universal service provision remains, however. In view of the special treatment it gets, it may be argued that SLTL also should bear greater responsibility for universal service provision. In any case the experience in Sri Lanka may confirm, as it has in many other countries, that competition may produce better universal service outcomes than monopolies.[13]

4.4 Tariffs

The government's 1994 national telecommunications policy and WTO agreements commit it to providing access to telecommunications facilities at cost-based prices. But again, in the absence of sound cost data, implementing that policy through regulation is not easy.

At present, no formal principle on tariff regulation is applied in the telecommunications sector. Whilst the licences given to SLTL and to the WLL operators contain provisions on price-cap regulation, they are currently on hold and may not be imposed until the year 2005 if they meet certain performance criteria (Samarajiva 1998). The government decision to re-balance SLTL rates over a five-year period, effectively suspends price-cap regulation for SLTL for that period. And while the licences given to the mobile and internet operators contain price-cap language, light-handed regulation is practised in these segments. The other segments of the telecommunications industry are subject to price-cap tariff regulation.

The official justification for tariff re-balancing mirrors the ITUs position that the practice of cross-subsidising the domestic market using net settlement payments from abroad should be abandoned (Utsumi 2000). A unilateral decision by the US telecommunications regulatory authority, the FCC, to establish lower benchmarks on settlement rates also impacts on developing countries like Sri Lanka, which have long benefited from the transfer of financial resources from developed countries under the current international telecommunication settlement rate system. According to the FCC transition schedule, Sri Lanka had to reach the benchmark rate of US$0.23 by 1 January 2002. Currently, US carriers pay SLTL a settlement rate of approximately US$1 per minute for terminating calls originating in the US.

SLTL reduced international tariffs by 8 per cent in 1998, and reached an agreement with the government to increase domestic tariffs to expand its domestic revenue base in five stages between 1998 and 2002: 1998 and 1999: 25 per cent; 2000: 20 per cent; 2001 and 2002: 15 per cent. Whether these increases are justified on a cost basis is unclear because of the absence of sound cost data. But the possibility that a dominant (and privileged) operator may use tariff re-balancing to charge anti-competitive 'quasi-monopoly' rates in the domestic market cannot be ruled out. Indeed, SLTL certainly extracted significant rents from the international segment (see Antelope Consulting 1998). With that source of rents now drying up, however, and fears that rival operators such as Suntel (which are linked to large multinational companies) may divert substantial levels of international traffic increasing, the incentives for SLTL to exploit the domestic market more intensively are obvious.

But what are the implications for the majority of users? The WLL operators appear to follow SLTL as the market leader in tariff setting. This, together with the high and stable domestic rates, suggests tacit price collusion amongst operators in the fixed market. Although some price flexibility has been introduced (for example, peak and off-peak prices; unbundling of customer premises equipment from installation and rental charges), overall rates remain high – given that the middle one-third of users pays between US$5.70 and US$14.30 per month for local and national calls, and that the lowest one-third pays between US$2.60 and US$5.70 in a country where around 21 per cent of the population lives below US$12.6 per person per month. TRC has recognised that if domestic prices continue to increase, affordability may become a serious problem for lower income subscribers. Thus, it has introduced special tariff schemes to partially shield low and medium users from tariff increases during the first two years of rate re-balancing. Given that the vast majority of fixed-access customers subscribe to domestic call services, and only a small proportion have access to international direct dialling facilities, there can be little doubt that rate re-balancing will have a negative impact on the welfare of most consumers. According to industry sources, in 1999, SLTL's international sector penetration was a mere 2.28 IDD lines per 1000 persons and the WLLs' subscriber base showed that only approximately 30 per cent of their clientele had IDD facilities.

4.5 The Political Economy of Anti-Competition

These issues highlight the fact that the movement towards a more competitive market is not a smooth one. There has been a discernible shift in the government's attitude towards competition and its enthusiasm for fostering competition appears to have waned. Indeed, some of the government's recent initiatives have a clear anti-competitive thrust. But what can explain the shift in government attitude from a pro-competition and privatisation stance to one of 'anti-competition'? We believe that the answer lies in the existing opportunities for rent extraction for politicians and favoured bureaucrats.

In an earlier analysis of the telecommunications industry (Jayasuriya and Knight-John 1998), we linked the early pro-competitive phase of the 1980s to the political inability of the government at the time to make fundamental changes to the existing state dominance. The strength of the trade unions in the telecommunications sector was one factor, but another was the prevailing ideological paradigm placed in the services such as telecommunications exclusively within the public sector. Opening up of new sectors (mobile, paging services) to new entrants was politically favourable, in line with the pro-private sector bias in policy, and a profitable source of rent extraction from licence seekers. But this changed when privatisation became a realistic option. Privatisation of state assets, particularly large and valuable public utilities, offered the possibility for rent extraction on a scale unimaginable in the past. This had been demonstrated during the last years of the previous United National Party (UNP) administration, albeit in a more limited scale. The politicians in opposition at the time (and their favoured friends and supporters) learned this lesson well. Thus, the current government's shift to a pro-privatisation strategy after assuming power is not difficult to understand. But privatisation offered even greater rents to those controlling the process if the enterprises to be privatised could be guaranteed monopoly profits. It helped that such policies could be lent legitimacy on the basis of maximising revenues from the sale of state assets; buyers of state assets would pay more if future profits were higher, and obviously they would be higher if monopoly power could be entrenched in lucrative markets. From the viewpoint of politicians and bureaucrats dealing with potential investors, regulatory restrictions that impeded future competitive threats was a policy that complemented privatisation. The less transparent the processes of privatisation, and the closer the ties between regulatory agencies and those in charge of the privatisation process, the greater the opportunity to maximise rents to politicians and new investors at the expense of public interest.

It is often argued that privatisation of state assets offers only a one-off opportunity for rent extraction to those with state control. Once privatised, that opportunity disappears, so the onetime 'loss' is worthwhile as a more efficient industry now emerges. But this is not always true. When what is being privatised is a public utility that once was a state monopoly, the buyer of the incumbent monopoly has an incentive to avoid pro-competitive regulations. The regulatory structure and the composition of the regulatory body itself become critical to the evolution of competition in the industry. The regulatory body can either aggressively pursue pro-competitive policies, or it can entrench the market power of the new private firm. If those who use their control over the state and the privatisation process to extract rents also can control the composition and mandate of the regulatory body, there is a double dividend from the process. Privatisation, therefore, is all the more profitable for politicians if it is followed by anti-competitive regulations.

This suggests a motive for, and an explanation of, the reasons why governments that appear committed to pursuing quite radical market-friendly policies sometimes shift their position from pro- to anti-competition in privatised industries. A control of the levers of the regulatory processes ensures that rent extraction can be maintained. While we do not suggest that this political economy logic explains the policy shifts in telecommunications in Sri Lanka, it does seem quite consistent with such an interpretation. To that extent it provides an illustration of the general point we seek to make regarding the link between rent-seeking behaviour, privatisation and anti-competitive regulatory policies.

5. CONCLUSIONS

International experience has shown that effective, and at times aggressively assertive, regulatory interventions are necessary to foster the competition required for the creation of an efficient, dynamic, and technologically innovative telecommunications industry that delivers cheaper and better quality services to consumers. This was the rationale for the setting up of the TRC. As stipulated in the provisions of the 1996 Act, the Commission has to foster effective competition in the industry while also accommodating the interests of a wide spectrum of stakeholders. Our analysis of the structure, conduct and performance of Sri Lanka's telecommunications industry suggests that some very crucial regulatory policy challenges need to be urgently addressed if the TRC is to become a credible and effective body. There is a widespread perception, not entirely unjustified, that the regulatory balance has tilted in favour of the dominant operator, SLTL.

Whilst the regulatory task of balancing conflicting stakeholder interests is difficult, even under the best of circumstances, the politicisation of the regulatory process makes the regulatory authority more vulnerable to capture by particular players in the industry and casts doubts on its impartiality. This makes it more difficult to attract investment into the sector, sends negative signals to the entire investor community, and undermines its capacity to play its full role in the economic development process. Ensuring the independence of the regulatory body and transparency of its procedures is the central political task facing the government, if its claims to uphold public interest are to be credible. It is time for the government/TRC to move from managing the industry in order to insulate the dominant, politically favoured operator from competitive pressures, and place its focus instead on enhancing market competition, improving consumer welfare and stimulating dynamic growth.

ACKNOWLEDGEMENT

The authors gratefully acknowledge the assistance of many people in the telecommunications industry. In particular they thank Professor Rohan Samarajiva (who headed the Telecommunications Regulatory Authority until 1999) for comments, assistance and encouragement, and representatives of the Telecommunications Regulatory Commission and the Public Enterprise Reform Commission, and several service providers for very useful discussions. However, responsibility for the views expressed and any remaining errors in this chapter are the sole responsibility of the authors. Authors' names are in alphabetical order.

NOTES

1. See Jayasuriya and Knight-John (1998) for a more detailed account of the liberalisation process and the political and ideological background to economic reform.
2. See Athukorala and Jayasuriya (1994) for a discussion of the 1977 reform process and investment programmes.
3. Profitability of telecommunications industries in developing countries have been generally higher than in developed countries (see Jayasuriya et al. (1997)) for a discussion of the 'investment paradox' in telecommunications).
4. The TRC itself has neither formally defined what constitutes market power nor made a formal determination involving the concept of market power.
5. Estimated on the basis of data from TRC and industry sources.
6. This issue relates to the controversy in the industry over enhanced voice operations and the alleged infringement of SLTL's monopoly rights in international telephony, and is discussed later.
7. Incidentally, this was the first public hearing initiated by the regulatory authority, despite the fact that the provision for public input into telecommunications regulation has been in place since 1991.
8. Calculated using figures cited in *The Daily News*, 19 August 2000.
9. The numbering system that currently exists in the Sri Lankan telecommunications market, distinguishes operators by allocating particular prefixes and a particular number of digits to a given provider. Attempts by the TRC to adopt a closed numbering plan have been stalled by SLTL.
10. See http://www.trc.gov.lk/onews.html
11. This operator also claimed to have plans to seek legal redress against the Sri Lankan government at the United Nations Human Rights Commission, New York, for arresting one of the firm's engineers under the Prevention of Terrorism Act on a complaint made by SLTL.
12. These statistics are for 1999 and come from personal communications with SLTL, WILL, and mobile phone sector sources.
13. See Wellenius (2000) for a discussion of universal service in competitive environments.

ABBREVIATIONS

BOO	build operate and own
BOT	build operate and transfer
CPP	calling party pays
DOT	Department of Telecommunications
EU	European Union
FCC	Federal Communications Commission
IDD	international direct dialling
IPO	initial public offering
ISP	internet service provider
ITU	International Telecommunications Union
MPP	mobile party pays
NTT	Nippon Telegraph and Telephone
ODGT	Office of the Director General of Telecommunications
PTT	post, telegraph and telephone
SLT	Sri Lanka Telecom
SLTL	Sri Lanka Telecom Ltd
SLTS	Sri Lanka Telecom Services Ltd
TRC	Telecommunications Regulatory Commission
UNP	United National Party
USO	universal service obligations
WAP	wireless access protocol
WLLs	wireless loop operators
WTO	World Trade Organization

REFERENCES

Antelope Consulting (1998), *Case Study of the Impact of the Changing International Telecommunications Environment on Sri Lanka*. A study by Antelope Consulting for the ITU and the Commonwealth Telecommunications Organisation, Colombo.

Athukorela, P. and S. Jayasuriya (1994), *Macroeconomic Policies, Crises and Growth in Sri Lanka, 1969–90*, Washington DC: World Bank.

Central Bank of Sri Lanka (various years), *Annual Report*, Colombo: Central Bank of Sri Lanka.

Duff & Phelps Credit Rating Lanka Ltd (2000), *Sri Lanka Telecom: Credit Analysis*. Colombo: DCR Lanka.

Government of Sri Lanka (1995), Policy Statement by Her Excellence Chandrika Bandaranaike Kumaratunga, President of the Democratic Socialist Republic of Sri Lanka, Colombo: Department of Government Printing.

Gunawardene, P. (1999a), 'Interconnection: the Sri Lankan experience', *Journal of the Asia Pacific Telecommunity*, **2** (1), 16–19.

Gunawardene, P. (1999b), *Telecommunications Access for Rural Sri Lanka*, (http://www.infocoms.com).

Gunawardene, P. (2000), 'Quality of service in telecoms: a perspective from Sri Lanka', paper presented at the Core Training Forum for Infrastructure Regulation (SAFIR), 10 February 2000, Agra, India.

ITU (International Telecommunication Union) (1999a), *World Telecommunication Development Report 1999: Mobile Cellular*. Geneva: ITU.

ITU (International Telecommunication Union) (1999b), *Trends in Telecommunication Reform 1999: Convergence and Regulation*, Geneva: ITU.

Jayasuriya, S. and M. Knight-John (1998), 'Liberalisation and privatisation: Telecommunications reforms', in Sri Lanka, in S. MacDonald and G. Madden (eds), *Telecommunications and Socio-Economic Development*, Amsterdam: Elsevier Science, pp. 353–73.

Jayasuriya, S., R. Maddock and R. Tourky (1997), 'Deregulation and the investment paradox in telecommunications', mimeo, La Trobe University, Melbourne: School of Business.

Knight-John, M. (1997), 'Private participation in infrastructure development', in *Infrastructure Development in Sri Lanka: Regulation, Policy and Finance*, Hong Kong: Euromoney Publications (Jersey), pp. 25–50.

Noll, R. (forthcoming), 'Telecommunications reform in developing countries', in A.O. Krueger (ed.) *Economic Policy Reform: The Second Stage*, Chicago: University of Chicago Press.

Samarajiva, R. (1998), 'Telecommunications policy and regulation in Sri Lanka', paper presented at TELEWORK-S and the South Asian Telecommunication Regulators' Council Meetings, 29 April–2 May 1998, Colombo, Sri Lanka.

Samarajiva, R. (1999), 'Sri Lanka's WTO implementation and universal access', *Connect-World Asia*, 1st Quarter, 11–14, (http://www.connect-world.co.uk/docs/articles/cwasia1q99/ samarajivasrilankacwasia1q99.asp).

Samarajiva, R. (2000a), *Communicating Legitimacy: Key to Effective Regulation,* Graduate Colloquium Series, School of Public Policy and Management, Ohio State University.

Samarajiva, R. (2000b), 'The role of competition in institutional reform of telecommunications: lessons from Sri Lanka', *Telecommunications Policy*, **24** (8/9), 699–718.

Utsumi, Y. (2000), 'Moving beyond international accounting rates', *Telecommunications Policy Online*, **24** (1), (http://www.tpeditor.com/utsumi.html).

Wellenius, B. (2000), 'Extending telecommunications beyond the market: toward universal service in competitive environments', *Public Policy for the Private Sector*, Washington DC: World Bank.

9. Vietnam: Restructuring telecommunications industry

Duc-Tho Nguyen, Quoc-Phuong Le and Nhat-Le Tran

1. INTRODUCTION

Since 1986, Vietnam has carried out a wide range of liberalisations and policy reforms, often known as *Doi Moi* (Renovation or Reform). As part of this process, changes have accelerated in the telecommunications industry, particularly during the 1990s. At the beginning of the decade, the industry was very small and under-developed. As of 1990, there were only about 100 000 telephone subscribers in the whole country, most of whom were officials in government and state organisations or state-owned businesses. The teledensity ratio was approximately 0.14 fixed telephone lines per 100 people, one of the lowest in the world (Nguyen 1999; DGPT and JICA 1999). The industry operated under strict state control, and there was effectively only one service provider, the state-owned Vietnam Posts and Telecommunications Corporation (VNPT).

The authorities recognised that the industry occupies a strategic role, in that it is a key component of the infrastructure required for national economic development. Accordingly, they have made substantial investments from the state budget, and have also gradually eased control over the industry, with the aim of rapidly expanding and upgrading capacity. By the year 2000, the number of fixed telephone lines had reached 3 million, raising the teledensity ratio to around 4 lines per 100 people (see Figure 9.1). Growth in other aspects of telecommunication usage has been similarly rapid.

Nevertheless, it should be kept in mind that in Vietnam, as in a number of other developing and transition economies, this industry is regarded as not only an important economic industry, but also a national asset, and a sensitive area where national sovereignty and security may be at stake. Accordingly, the official approach to liberalisation has been gradual and cautious, and this will probably not change very much in the near term.

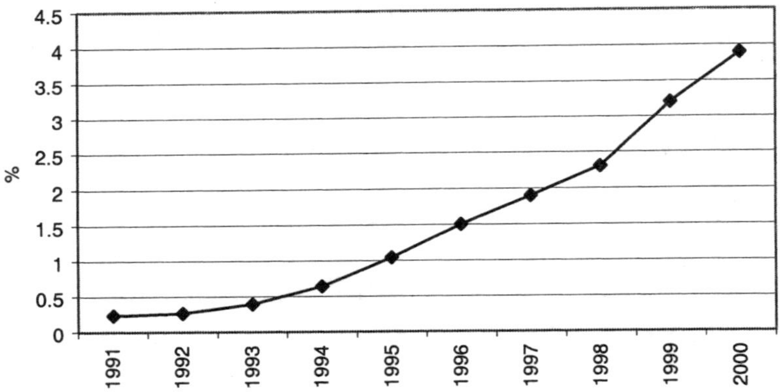

Figure 9.1 Teledensity in Vietnam, 1991–2000

In this chapter we present a brief review of the main regulatory changes and other key developments in the telecommunications industry in Vietnam, and an assessment of the industry's achievements during the past decade, as well as some of the major challenges facing it in the near future.

2. REGULATORY AND OPERATING ENVIRONMENT

2.1 Regulatory Framework

The following legal documents are the main components of the industry's regulatory framework:

- Government Decree 109/1997/ND-CP (12 November 1997) on Posts and Telecommunications;
- Ordinance on Posts and Telecommunications, promulgated on 11 June 2002;
- Government Decision 158/2001/QD-TTg (8 February 2001) on the 2001–2005 Development Plan for the Posts and Telecommunications sector, updating a similar Decision issued in February 1997;
- Government Decision 81/2001/QD-TTg (24 May 2001) ratifying the Action Plan to carry out Politbureau Directive 58/CT/TW on promoting the utilisation and development of telecommunication technology in industrialisation and modernisation during 2001–2005;

- Government Decree 55/2001/ND-CP (23 August 2001) on the regulation, provision and utilisation of the internet in Vietnam, replacing a similar Decree issued in March 1997;
- The Law on Foreign Direct Investment, the Law on Labour, and other related laws and regulations.

Historically, postal and telecommunications services in Vietnam have always been administered together (this practice dates back to before independence). At present, both types of services are regulated by the Department-General of Posts and Telecommunications (DGPT), whose responsibilities also encompass the management of radio frequency. In the Vietnamese bureaucracy, DGPT's rank is deemed equivalent to that of a Ministry. The government (at Cabinet level) determines broad policies governing the industry, and DGPT provides guidelines as to how these policies are to be interpreted and implemented in practice, and supervises compliance (see Figure 9.2). The new Ordinance on Posts and Telecommunications, promulgated on 11 June 2002, clearly indicated the authorities' intention to separate the *operations* of postal and telecommunications services. Nevertheless, to date there has been no announced plan to separate the *supervisory* functions of DGPT along similar lines.

2.2 Main Operator

The dominant operator in the telecommunications industry is VNPT, a giant state-owned conglomerate of companies. Its rather complicated organisational structure as at the beginning of 2001 is illustrated in Figure 9.2. Besides the functional divisions at the head office and a number of auxiliary units (such as training centres) which may be located elsewhere, VNPT's sub-units also included the local posts and telecommunications (P&T) offices in the country's 61 provinces, and a range of subsidiaries or 'member companies'. Some of the latter (such as Vietnam Telecommunication National, Vietnam Telecommunication International, Vietnam Data Company, Vietnam Posts Services Company, and Posts Savings Company) are financially not distinct from VNPT as a whole, while others (such as Vietnam Philately Company, Vietnam Mobile Communication Company, P&T Financing Company, and a number of design and construction companies) are largely independent from VNPT in terms of financial management. Some of these 'financially independent' member companies have also been 'equitised' (that is, partially privatised).

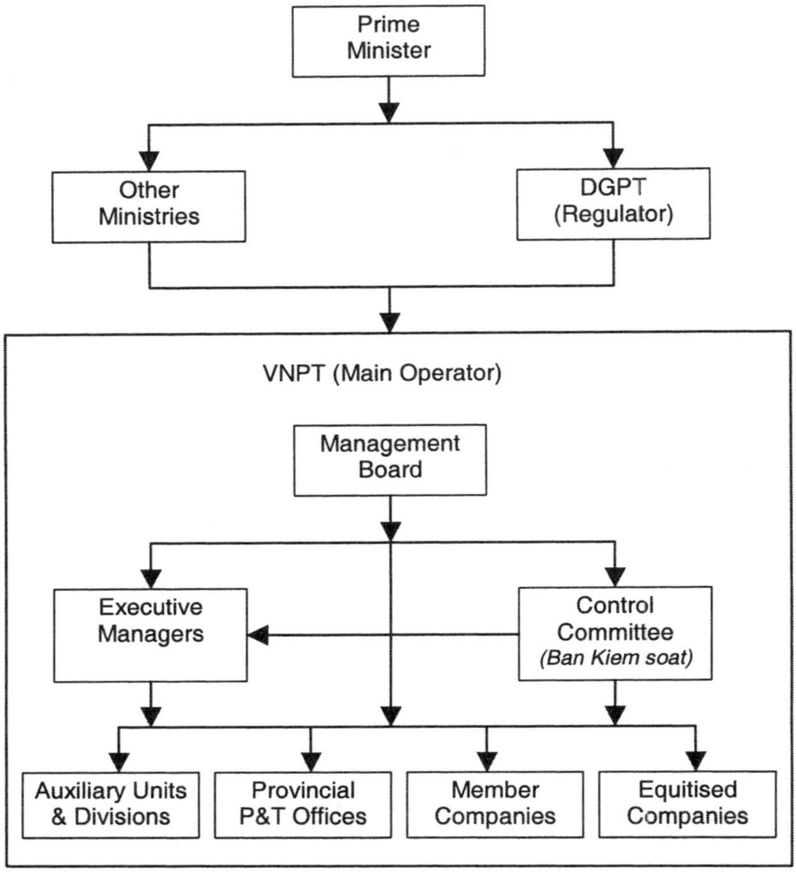

Source: VNPT Website

Figure 9.2 Regulatory and operating structure of Vietnam's telecommunications

Previously the functionalities of both DGPT and VNPT were vested in the same state agency and there was no effective separation of regulatory and operational responsibilities. This situation ended in 1990 when VNPT was established as a state-owned business and the regulatory function was entrusted initially to the Posts and Telecommunications Department within the Ministry of Transportation (from 1990 to 1992) and then DGPT (from November 1992 to date). Despite this crucial separation, and a number of other important regulatory changes which will be examined below, VNPT has continued to dominate the industry.

A major restructuring of VNPT began in the second half of 2002. The new Ordinance on Posts and Telecommunications envisaged that there would be a separate and financially independent corporation, called Vietnam Posts Company, which would inherit from VNPT the provision of postal services nationwide. Accordingly, VNPT announced plans to complete this corporate separation process by the year 2006 (*Sai Gon Giai Phong*, 30 May 2002). Indeed, as early as September 2001, the corporation had begun to carry out, on an experimental basis, the separation of postal and telecommunications services in ten provinces and major cities. Nationwide separation of the two types of services began in earnest on 1 July 2002. In the remainder of the chapter, we shall restrict our focus to the telecommunications industry only.

3. POLICY AND REGULATORY CHANGES

3.1 Overview

In broad terms, policy and regulatory changes have occurred in four main areas:

1. Since 1988, foreign companies have been allowed to either establish local operations to produce telecommunication equipment and material on their own, or assist local operators (mainly VNPT) in the provision of services;
2. Since 1995, new local companies have been allowed to provide telecommunication services, thus ending VNPT's virtual monopoly, and new major services have been introduced;
3. Since 1998, a number of state-owned telecommunication companies, including VNPT, have begun an 'equitisation' process which increases the private sector's involvement in the provision of capital for further investment in the industry; and
4. Since 1997, service providers have been allowed greater flexibility in setting prices, thus laying the foundation for future price competition, and the authorities have made efforts to render regulations more transparent and streamlined.

Each of these areas will be examined in turn below.

3.2 Foreign Participation

Foreign investment in the telecommunications sector has been welcome for at least two reasons. First, the amount of capital required to finance the much-needed capacity expansion far exceeds the locally available funding, which comes mainly from the state budget. Second, foreign technology is attractive in that it offers the promise of being able to quickly upgrade local systems through a 'catch-up' process which bypasses outdated technology and directly accesses latest innovations.

In 1988, the first foreign direct investment (FDI) project in this industry was established by VNPT and Telstra (Australia). By 1998, a total of 24 FDI projects had been approved (see Table 9.1). These include:

- twelve business-cooperation contract (BCC) projects;
- eleven joint ventures (JV); and
- one 100 per cent foreign-invested enterprise (FIE).

To date, all FDI projects which are related to the provision of telecommunication services must be in the form of BCCs. They tend to be fairly substantial, with the biggest project being worth US$467 million. The total value of capital committed in these projects to 1998 was around US$1,100 million.

A BCC is a profit-sharing contract between a foreign investor and a Vietnamese partner, which does not require the establishment of a new legal entity. The foreign partner usually trains personnel and provides technical consultation, while the Vietnamese partner (most often VNPT, or one of its financially autonomous member companies) provides services to the customers and then shares any earned profits with the foreign partner. The latter has no direct access to the customers, little or no control over pricing, and less influence over profit determination than under the other FDI forms, such as JV, 100 per cent FIE, or build–operate–transfer (BOT).

These alternative forms of FDI have been allowed in the case of foreign companies which do business in any aspects of telecommunications except services provision. The relevant projects have been relatively small, ranging from $0.3 million to $25 million, and their total value to 1998 was only about $100 million despite the fact that they were more numerous than the BCC projects. The extension to these FDI forms of approval for foreign investment in service-provision related activities has been under consideration for some time. Nevertheless, as of 2000, BCC remained the only allowable form of FDI for foreign investors who wish to take part in providing telecommunication services. Undoubtedly this restriction has had some dampening effect on the level of FDI in the industry.

It is expected that policy in this area will be changed dramatically in the next few years following the signing in July 2000 of the Vietnam–US Bilateral Trade Agreement (BTA). Some of the changes which are likely to be implemented are examined in section 5.2 below.

Table 9.1 FDI projects in the telecommunications industry

	Foreign partner	Vietnamese partner	Investment form	Activity	Duration	Investment capital ($US million)	Status at writing
1	Telstra, Australia (69%)	VNPT (31%)	BCC	Telecommunication network expansion	12 yrs (1988–2000)	237	Expired
2	France Cable et Radio Vietnam (100%)	VNPT (0%)	BCC	Upgrade telecommunication network, supply services	15 years	467	Active
3	NTT Vietnam, Japan (100%)	VNPT (0%)	BCC	Upgrade telecommunication network, supply services	15 years	194	Active
4	Sapura, Malaysia (62%)	HCMC Post (38%)	BCC	Public card phone	8 years (1993–2001)	5.2	Active
5	Cable & Wireless, UK 100%)	VNPT (0%)	BCC	Exploit local network	15 years (1998–2013)	207	Active
6	Voice International, Australia (70%)	Intl Comm Service (30%)	BCC	Telecommunication network for HCMC	1989–	0.73	Active
7	OTC, Australia (69%)	VNPT (31%)	BCC	Construct satellite station	10 years (1988–98)	130	Expired
8	Cable & Wireless Vietnam, UK (99%)	VNPT (1%)	BCC	Upgrade telecommunication network	15 years (1998–2013)	207	Active
9	OTC, Australia (90%)	VNPT (10%)	BCC	Set up 2 satellite stations	10 years (1990–2000)	287	Active
10	Korea Telecommunication (75%)	VNPT (25%)	BCC	Develop telecommunication network in Hai Phong, Hai Hung, Quang Ninh	10 years (1996–2006)	53	Active

Table 9.1 (continued)

	Foreign partner	Vietnamese partner	Investment form	Activity	Duration	Investment capital ($US million)	Status at writing
11	Worldcorp, Singapore (58%)	VNPT (42%)	BCC	Telephone directories	5 years (1995–2000)	0.38	
12	Comvik Intl, Sweden (47%)	Mobile Com (53%)	BCC	Develop mobile phone network	10 years (1995–2005)	341.5	Active
13	Alcatel, France (51%)	VNPT (49%)	JV	Manufacture, install, operate services	10 years (1993–2003)	4.65	Active
14	Viiti Investment, Hong Kong (60%)	HCMC Post (40%)	JV	Telephone, telex, fax service	5 years (1991–96)	0.3	Expired
15	NEC, Japan (51%)	VNPT (49%)	JV	Manufacture digital electronic directory systems	5 years (1997–2002)	15	Active
16	Fujitsu, Japan (50%)	VNPT (50%)	JV	Produce, install telecommunication systems	15 years (1997–2012)	12	Active
17	Siemens, Germany (51%)	VNPT (49%)	JV	Manufacture fibre optic cable	20 years 1994–2014)	14.9	Active
18	Harrison Industries, US (57%)	Phuong Dong Co. (43%)	JV	Manufacture telecommunication and electronic products	30 years (1997–2027)	25	Active
19	Sun Moon Star, Hong Kong (69%)	TRAMATSUCO (31%)	JV	Assemble telecommunication equipment	20 years (1990–2010)	1.5	Active
20	Newtel, US (70%)	SIGELCO* (30%)	JV	Manufacture electronic and telecommunication equipment	40 years (1995–2035)	20	Active

Table 9.1 (continued)

	Foreign partner	Vietnamese partner	Investment form	Activity	Duration	Investment capital ($US million)	Status at writing
21	Siemens, Germany and Taiwan (60%)	VNPT (40%)	JV	Manufacture, install EWSD station	15 years (1995–2010)	5	Active
22	Viba, overseas Vietnamese in Germany (30%)	COKYVINA (70%)	JV	Manufacture telecommunication products	5 years (1988–93)	0.5	Expired
23	Yeebo Development, Hong Kong (60%)	HCMC Post (40%)	JV	Telephone manufacture	10 years (1991–2001)	0.9	Active
24	Harada Industry, Japan (100%)		100% foreign	Manufacture car antennae	50 years (1997–2047)	6	Active

Note: * Sigelco is the former name of Vietel

Source: Compiled from Vietnam Online (http://www.vietnamonline.net)

3.3 Entry of New Local Service Providers and Introduction of New Major Services

Before 1995, VNPT was the only company allowed to provide telecommunications services to the general public. In 1995, the authorities took the first step toward allowing greater competition in this industry by allowing two other domestic companies to provide telecommunications services: the Military Electronics and Telecom Company (Vietel), and Saigon Joint-stock Posts and Telecommunications Corporation (Saigon Postel, or SPT), both of which were established in that year.

Despite this step, VNPT faced little effective competition until internet services were officially introduced in March 1997 with the issuance of Government Decree 21/CP. In November 1997, five local companies were licensed to become internet service providers (ISPs). These included VNPT, Vietel, SPT, and two other players: Financing and Promoting Technology (FPT) and Netnam. (As at writing, Vietel has not actually utilised this opportunity to provide internet services.)

The ISP sub-market has been one of the fastest-growing segments of the entire telecommunications industry. Nevertheless, reflecting the fact that the internet is considered a sensitive area which could affect national culture and security, the authorities have reserved the right to control the flow of information, and to block any information which it regards as harmful, on the grounds that such literature may incite hatred and violence, aggravate conflicts between nationalities and countries, or promote social vices.

As at the end of 2000, VNPT still held the lion's share (about 62 per cent) of the ISP market segment. Indeed, even after the considerable changes which had occurred during the second half of the 1990s, at the end of the decade VNPT remained dominant both overall and in most segments of the telecommunications market. The only exception to this at the time was the ship-to-shore and shore-to-ship sub-market, and the Inmarsat services sub-market, in which Vietnam Shipping Communications and Electronics Company (Vishipel) dominated. Vishipel, which was owned by the Ministry of Transport, was allowed to establish and control all coastal communication stations (there were eight such stations at preparing this chapter) as well as the country's only Inmarsat Land-Earth Station.

At the end of the 1990s, VNPT – through its subsidiary Vietnam Data Company (VDC) – was still effectively the only operator licensed to provide internet exchange points (IXP). In 2002, DGPT began to allow FPT and Vietel to provide IXP services. According to official plans, there will be 3–5 IXPs and 30–40 ISPs by the year 2005.

Competition has also become more vigorous in other market segments during recent years. At the beginning of 2001, Vietel was licensed to provide domestic long-distance services between Hanoi and Ho Chi Minh City on a trial basis. In 2002, S-Telecom, a joint venture between SPT and three Korean companies (SK Telecom, LG Electronics, and Dong Ah Elecomm Inc) unveiled a project to deploy a nationwide mobile network based on the CDMA technology.

In March 2002, VNPT began to face competition in the sub-market for local fixed-line voice telephone services, as Saigon Postel gained permission to provide such services in the nation's busiest commercial centre, Ho Chi Minh City, thus becoming only the second operator, after VNPT, to have been allowed to provide these services anywhere in Vietnam. Saigon Postel has had to negotiate with VNPT to integrate with the latter's fixed-line voice telephone network (VASC 2002).

Another potentially formidable competitor that has emerged recently is the Electric Telecommunications Company (ETC), an affiliate of Electricity of Vietnam (EVN), the country's sole supplier of electricity. In June 2002, ETC gained permission to provide local and distance fixed-line telephone services, in addition to its existing ISP licence. ETC enjoys at least two special advantages compared with other telecommunications companies. First, it can share the nationwide fee collection network of EVN, its parent company, thereby reducing administration costs. Second, it can potentially develop its own long-distance fixed-line infrastructure which would capitalise on EVN's assets associated with the nationwide electric cable network (Khoa 2002).

3.4 Partial Privatisation

Although the telecommunications market has been opened up to additional service providers since the mid-1990s, government regulations have continued to limit eligibility to play such a role to entities which are either fully state-owned or joint-stock companies in which the state holds a 'dominant or special share' (Decree 109/1997/ND-CP). This is a significant regulation, especially in view of the country's broader agenda of partially privatising many state-owned enterprises through an 'equitisation' programme, which has been implemented since 1998 following Government Decree No 44/1998. Under this programme, selected fully state-owned companies would become joint-stock, with parts of the state's shareholdings being offered to private businesses and individuals.

As of 2000, VNPT itself remained fully state-owned, as were Vietel, FPT, Netnam and Vishipel. By contrast, SPT was a joint-stock company. The relationships between SPT and VNPT have been particularly interesting. SPT was the first provincial/municipal Posts and Telecommunications Company to become joint-stock (it did so in 1995). SPT is also an independent services provider and does not act as a VNPT's subsidiary nor local company in Ho Chi Minh City. As of 1999, 90 per cent of its equity remained in the hands of state-owned enterprises, with VNPT owning about 18 per cent. Further, for at least one historical episode, the Chair of SPT's Management Board was at the same time a member of VNPT's Board (Levine 1997).

While VNPT itself has remained fully state-owned, it has begun to partially privatise some of its subsidiaries during the last few years. Cables and Telecommunication Materials Company (Sacom), one of VNPT's affiliates, was one of the first two joint-stock companies to be listed on Vietnam Stock Exchange when it opened in July 2000. (The other company was Refrigeration Electrical Engineering Company, or REE.) Since then, Sacom has always been among the most attractive shares being traded in the stock market. This positive outcome has encouraged VNPT to extend the equitisation programme to other affiliates (Ly 2002c).

During its annual meeting held in February 2002, VNPT unveiled plans to restructure itself into a multi-sectoral, joint-stock conglomerate operating in information technology as well as telecommunications. The new conglomerate, which will officially come into being by 2005, will be a mixture of state and private ownership, with the State holding a majority interest (Sam 2002).

Potentially the equitisation programme can lead to a much greater role for the private sector in financing future investment and growth in the economy generally. However, with the restrictions on foreign investment in the telecommunications sector still in place (as outlined in section 3.2) and with VNPT continuing to play such a dominant role, it will be some time before this potential benefit can be fully realised in the context of this particular sector.

3.5 Greater Flexibility in Pricing and More Transparent Regulations

Before 1997, all prices in the industry were set by the government (Government Decree 137/1992/ QD-HDBT). In 1997 this area became subject to partial deregulation. As stipulated by Government Decree 109/1997/ND-CP issued on 12 November 1997, there is a hierarchy to be followed in authorising telecommunications prices:

- The government now sets the pricing framework and individual prices for *key public* services only, such as the charge per minute of local calls;
- DGPT sets the pricing framework and individual prices for the *remaining public* services, for example tariffs relating to leased-lines services; and

- Services providers are free to set prices for specific services within the above frameworks and for other services, for example charges for value-added services such as call forwarding or call waiting.

As an illustration of this pricing regime, consider the monthly rental fee for the basic telephone service and the corresponding installation fee. As at 2000, the monthly rental fee for fixed-line telephone service was set as an 'individual' (that is, fixed) price, which was VND30,000 (equal to US$2.10) in urban areas and VND20,000 (US$1.40) in remote areas. All service providers were obliged to charge the above fixed fees nationwide. By contrast, the once-only installation fee was set as a 'ceiling' price, which was VND1,500,000 (US$110) in urban areas and VND500,000 (US$37) in remote areas. Service providers were allowed to charge a fee lower than these ceilings, based on their own assessment of the geographical, economic and commercial circumstances.

In April 2002, DGPT conducted a ballot to gauge the reaction of local ISPs to two proposed options for deregulating internet service prices, namely (a) totally free-market prices and (b) floor prices to be set by DGPT. Under the first option, local ISPs would be free to set their own prices without any official interferences. Under the second option, service providers would have to set prices at least as high as the officially determined minimum. Interestingly, most local ISPs (with VNPT being a notable exception) opposed the first option. The smaller service providers argued that, as fair-competition laws and regulations were as yet relatively under-developed, big players such as VNPT and the multinationals would be able to engage in price dumping to drive the other operators out of this rapidly evolving market (Ly 2002a).

In Vietnam, the basic telephone service (as well as water and electricity) is typically seen as a 'public' service, not in the sense encapsulated in the economics concept of a 'public good', but in the everyday-usage sense that it is a service which is expected to be provided to all residents at an affordable price. In other words, it is seen as a service where universal service requirements are considered almost paramount. In order to achieve this objective, there has been considerable cross-subsidisation of non-business by business customers, and of basic by non-basic services – particularly of local, fixed-line calls by long-distance and international calls.

As the dominant operator, VNPT has played a major role in the delivery of universal services throughout the country. Accordingly, it has been argued that the interconnection fees which VNPT is allowed to charge other service providers should take account of the need for some cross-subsidisation to finance universal services, as well as the imperative to achieve greater allocative efficiency through transparent and consistent pricing.

In September 1998, DGPT issued a Decision on Interconnection Arrangements among PSTN Networks (Decision No 547/1998/QD-TCBD, dated 3 September 1998). The Decision sets guidelines for interconnection negotiations, prescribes the main terms and conditions to be adopted in interconnection arrangements, and also determines the procedures to be followed in dealing with interconnection disputes, in which DGPT would play the role of arbitrator.

In recent years, however, it became clear that interconnections between PSTN networks have not yet become the main issue. Instead, an area of pressing concern has been the interconnection of other types of services with VNPT's PSTN network, a case in point being the VoIP (Voice over internet Protocol) service which Vietel wanted to provide, with no extant systematic regulations to guide and govern the necessary negotiations. This example illustrates the tendency of technological innovations to outstrip regulatory changes, and the need for continuous updating and streamlining of regulations.

As part of its efforts to make telecommunications regulations more transparent, in recent years DGPT has begun to publish its new rules and regulations in the *National Gazette* (in both Vietnamese and English) as well as on a web-site reserved for legal matters (www.vietlaw.gov.vn). This is a welcome development, as it increases the flow of information among both existing and potential players, and helps to inform the public as well as key stakeholders of recent and proposed changes in the sector.

4. ACHIEVEMENTS

There can be no doubt that growth in coverage and total usage has been spectacular, albeit from a low base. As can be seen from Tables 9.2 and 9.3, while population growth during most of the 1990s averaged around 1.6 per cent per year, and real GDP growth was around 6–8 per cent per year, the number of fixed lines was doubling every two or so years, and the number of mobile phones almost doubled every year (*Posts and Telecommunications Statistical Compendium,* various issues; *Posts and Telecommunications Weekly*, various issues).

Similarly, the total duration of long-distance domestic and international calls doubled every two to three years. Growth in the less heavily regulated sub-market for internet services has been even more rapid. Since November 1997, when these services were first made officially available, the number of subscribers has increased exponentially, from only 1164 in 1997, to 18 743 in 1998, 45 000 in 1999 and about 100 000 at the end of 2000.

Table 9.2 Coverage and usage levels

	Population (million)	Fixed lines (000)	Lines per 100 persons	Mobile phones (000)	Domestic distance calls (m. minutes)	International calls (m. minutes)	Internet subscribers (000)
1993	69.6	255	0.4		235	64	
1994	70.8	443	0.6		449	137	
1995	72.0	734	1.0		760	208	
1996	73.2	1 110	1.5	71	1 130	273	
1997	74.3	1 398	1.9	147	1 442	335	1
1998	75.5	1 741	2.3	227			19
1999	76.6	2 419	3.2	308			45
2000	77.7	3 019	3.9	640			100

Source: Compiled from various sources

Table 9.3 Growth in coverage and usage levels (%)

	Population	Real GDP	Fixed lines	Mobile phones	Domestic distance calls	International calls	Internet subscribers
1994	1.7	8.8	73.9		91.1	112.1	
1995	1.7	9.5	65.9		69.3	51.9	
1996	1.6	9.3	51.2		48.7	31.6	
1997	1.6	8.1	25.9	107.6	27.6	22.7	
1998	1.5	5.8	24.5	54.4			1 510.2
1999	1.5	4.8	39.0	35.9			140.1
2000	1.4	6.7	24.8	107.7			122.2

Source: Compiled from various sources; Table 9.2

These statistics clearly demonstrate the system's successes in bringing existing and new services to a rapidly expanding customer base.

It is particularly encouraging to note the marked widening in the geographical and socio-economic coverage provided by the expanding and multiplying networks. For example, in September 2000 about 85 per cent of the country's communes had access to telephone services, compared to only 8 per cent in 1991.

5. REMAINING CHALLENGES

5.1 Continuing Need for Further Upgrading and Expansion of Capacity

By regional standards, Vietnam's telecommunications prices, especially those for both domestic and international long-distance calls, remain high. Since 1990, VNPT has reduced the *nominal* prices on international calls seven times, each time by about 10–15 per cent. After allowing for inflation, the price reductions in real terms have been very substantial indeed. Nevertheless, by VNPT's own estimates, prices in Vietnam remain some 25–50 per cent higher than elsewhere in the region (*Posts and Telecommunications News*, No. 4, January 2000).

As at the end of 2000, a 3-minute call from Ho Chi Minh City (HCMC) to Singapore cost around US$6.45. A 3-minute call between the two largest cities in the country, HCMC and Hanoi, cost VND11,800 (nearly US$0.80), and 10 minutes on the internet cost VND2,500 (nearly US$0.17).

As pointed out earlier, the installation fee for a fixed telephone line depends on geographical locations. In 2000 it ranged from about VND500,000 to VND1,500,000 (US$37 to 110). While these represented a considerable reduction (especially in real terms) from previous levels, they were still quite sizeable compared with per capita GDP (about US$296 in 1997). In October 2000, monthly rental costs were reduced from VND20,000 to 30,000 (around US$1.40 to 2.10) compared with a *combined* fee of VND68,000 before this date for both the monthly rental and 450 minutes of local calls.

The main reason for prices not coming down faster has been a continuing excess demand for services. While the average waiting period for fixed-line connections has fallen substantially during the past decade, it is still fairly long in many instances. The quality of access in rural areas remains poor. In turn, this shortfall between available supply and potential demand can be attributed to the fact that the amount of capital required to upgrade and expand services has far exceeded the funding available from the state budget and other official sources. Some of this funding gap has been met by official development assistance (ODA) but, in the longer term, much of it will have to come from the private sector, both foreign and domestic.

The need for upgraded capacity and new technology as well as financial depth gains added urgency when viewed against the global trends toward agglomeration and strategic alliances which allow the major players on both the global and regional scenes to exploit economies of scale and scope. For Vietnam to be able to develop a modern, up-to-date telecommunications industry, in the longer run it will be necessary to look beyond the national boundaries. Vietnamese carriers and service providers will need to develop true, level-playing-field competitiveness domestically in order to have any hope of being competitive in the region.

5.2 International Commitments to Liberalise Telecommunications

Vietnam joined the Association of South-East Asian Nations (ASEAN) in 1995 and the Asia Pacific Economic Cooperation (APEC) grouping in 1998. The country also signed a bilateral trade agreement with the USA in July 2000, and is seeking to join the World Trade Organization (WTO). Under the terms of relevant agreements and negotiations, Vietnam has pledged to further open its telecommunications industry to foreign investment and to further liberalise the industry's regulations.

For example, under the auspices of APEC, in 1998 Vietnam drew up an Individual Action Plan (APEC Secretariat Online). Among other things, the Plan envisaged that in the *short term* (1998 to 2000) the country would introduce new technologies and services such as Frame Relay and Voice over internet Protocol (VoIP); refine licensing procedures for equipment certification and service provision; and participate in regional telecommunication activities such as ASEAN MRA (mutual recognition agreement) and APEC MRA on equipment certification. In the longer term (2001 to 2020), Vietnam planned to promulgate a Law on Posts and Telecommunications; review existing policies to create a more favourable environment for both local and foreign participation; and look toward penetrating regional markets. An updated Action Plan was developed in 2002 against the timeframe of 2000–20.

Vietnam has also made commitments to ASEAN regarding telecommunication services, including electronic mail, voice mail, electronic data interchange, telex and telegraph. However, to date these commitments have not been very stringent, as they simply reiterate that while foreign companies are free to establish representative offices in Vietnam, these offices are not allowed to carry out any business activities, and that in order to provide services, foreign companies can cooperate with local companies under the BCC (Business Cooperation Contract) form.

Of greater interest have been the commitments made under the terms of the VN-US BTA. In particular, the following milestones were set for the opening of the telecommunication *services* sub-market to US investors:

- Two years after entry into force of the Agreement (three years in the case of internet services): joint venture enterprises will be permitted to provide *value-added* services, provided that the capital contribution of the foreign side may not exceed 50 per cent. Joint venture enterprises will not be permitted to construct their own long-distance and international circuits and will have to lease them from local operators;
- Four years after the Agreement: joint venture enterprises will be allowed to provide *basic* services, including circuit-switched data, packet-switched data, telex, telegraph, private leased circuit, radio-based services such as cellular mobile and satellite services. The capital contribution of the foreign side may not exceed 49 per cent. Joint venture enterprises will continue to be barred from constructing their own long-distance and international circuits.
- Six years after the Agreement: joint venture enterprises will be permitted to provide *telephone* services, including fixed local, long-distance and international services. The capital contribution of the foreign side may not exceed 49 per cent of the legal capital of the joint venture.

These provisions were consistent with the authorities' previously announced roadmap for foreign participation in telecommunication services, namely:

- To open the market for value-added network services (VANS) before opening the market for basic services;
- To open the market for local (mobile) calls before opening the markets for long-distance domestic and international calls; and
- To open the market for mobile services before opening the market for fixed voice services.

In order to meet another set of its APEC and US BTA commitments, as well as to prepare for WTO accession, since the late 1990s the authorities have worked to further reduce official control over prices in this sector. As has been seen in section 3.4 above, however, an immediate and complete deregulation of prices (for example, for internet services) may not necessarily be desirable: other regulatory mechanisms (for example, fair competition laws) may need to be in place first to prevent anti-competitive behaviours.

6. CONCLUSION

Since the early 1990s, major changes have occurred in Vietnam's telecommunications sector. Policy and regulatory liberalisations have played a key role, in that they have allowed foreign investors and domestic service providers to respond to both accumulated and growing demands for up-to-date services. As a result, the country has made significant strides towards catching up with regional and international standards.

Throughout this process, the authorities have been confronted with a dilemma. On the one hand, there has been an urgent need to rapidly expand and upgrade networks and services, which in turn implies a greater role for foreign and private domestic investors. On the other, the industry continues to be seen as vital to the country in non-economic as well as economic terms, such that the authorities have been determined to maintain a high level of effective control.

In the past, there has been so much scope for catching up – in that the country was starting from such a low base – that gradualism and caution have not prevented substantial progress from being made. By contrast, the industry is now poised far more delicately between further rapid development and relative stagnation (compared with regional standards). Opportunities for catching up exist elsewhere as well as in Vietnam, and the fact that foreign investment is a scarce commodity on a global scale well bears repetition.

ACKNOWLEDGEMENT

The authors wish to acknowledge the valuable research assistance provided by Quan Nguyen. The views expressed herein do not necessarily reflect those of the authors' respective employers.

ABBREVIATIONS

APEC	Asia Pacific Economic Cooperation
ASEAN	Association of South-East Asian Nations
BCC	business cooperation contract
BOT	build–operate–transfer
BTA	bilateral trade agreement
DGPT	Department General of Posts and Telecommunications
ETC	Electric Telecommunications Company
EVN	Electricity of Vietnam

FDI	foreign direct investment
FIE	foreign invested enterprise
FPT	Financing and Promoting Technology
IAP	internet access provider
ISP	internet service provider
IXP	internet exchange points
JV	joint venture
ODA	official development assistance
P&T	Posts and Telecommunications
SPT	Saigon Posts and Telecommunications Joint-Stock Corporation
VANS	value-added network services
VDC	Vietnam Data Company
Vietel	Vietnam Military Electronics and Telecommunications Company
Vishipel	Vietnam Electronics and Communications Shipping Company
VNPT	Vietnam Posts and Telecommunications Corporation
VoIP	voice over internet protocol
WTO	World Trade Organization

REFERENCES

APEC Secretariat Online, *Vietnam's Individual Action Plan 1998 under APEC*, (http://www.apecsec.org.sg)

Cohen, M. (2002), 'A tough call in Vietnam', *Far Eastern Economic Review*, **165** (17), 32–3.

DGPT (Department General of Posts and Telecommunications) and JICA (Japan International Cooperation Agency) (1999), *The Study on Telecommunications Development in the S.R. Vietnam*, August, Hanoi.

Foreign Investment Projects in Vietnam, Vietnam Online (http://www.vietnamonline.net)

Khoa, T. (2002), 'Cung cap dich vu qua he thong dien – Mot loi the kinh doanh moi cua ETC' [Electric cable network for telephone service – A new competitive advantage of ETC], *Vietnam Economic Times*, Hanoi.

Levine, J.J. (1997), 'Telecoms take-off', *Vietnam Business Journal*, Online, October.

Ly, N. (2002a), 'Cuoc Internet – Doanh nghiep khong muon coi troi' [Internet charge – Businesses do not want to be unshackled], online news service, VASC, http://www.vnn.vn/i-today/, accessed 29 June 2002.

Ly, N. (2002b), 'VNPT lua chon USTarcom cung cap he thong di dong noi vung' [VNPT selects USTarcom as provider of IP Personal Access System], online news service, VASC, http://www.i-today.com.vn, accessed 25 June 2002.

Ly, N. (2002c), 'VNPT no luc tham nhap thi truong chung khoan' [VNPT endeavours to enter the stock market], online news service, VASC, http://www.i-today.com.vn, accessed 25 June 2002.

Manyin, M.E. (2001), 'The Vietnam – US Bilateral Trade Agreement', CRS Report for Congress RL 30416, United States of America's Congressional Research Service.

N.A. (2000), 'Communist Vietnam's first stock exchange opens', online news service, BBC News, http://news.bbc.co.uk/hi/english/world/asia-pacific/newsid_855000/855846.stm, accessed 30 June 2002.

N.A. (2001), 'Vietnam's internet subscribers increase nearly three times', Xinhua News Agency.

N.A. (2002), 'Vietnam to reduce telephone charges', Xinhua News Agency.

N.A. (18 June 2002), 'CDMA Mobile Network will be available by the year end' [Online newspaper], Vietnam Economy Online, http://www.vneconomy.com.vn, accessed 19 June 2002

Nguyen, D. (1999), 'Vietnam's post and telecommunications prospects', in Van Hoa Tran, (ed.), *Sectoral Analysis of Trade, Investment and Business in Vietnam*, Basingtoke: MacMillan.

Posts and Telecommunications Statistical Compendium 1996–1998; Posts and Telecommunications Publishing House, August 2000, Hanoi.

Posts and Telecommunications Weekly, Online, various issues.

Sam, H. (2002), 'VNPT: Tong ket nam 2001 – Chien luoc cho nam 2002' [VNPT: Review of 2001 – Strategy for 2002], (Online news service), VASC Online, http://www.vnn.vn/i-today/, accessed 28 June 2002.

Tipton, F.B. (2002), 'Bridging the digital divide in Southeast Asia: Pilot agencies and policy implementation in Thailand, Malaysia, Vietnam, and the Philippines', *ASEAN Economic Bulletin*, **19** (1), 83–99.

Tran, Van Hoa (ed.) (1999), *Sectoral Analysis of Trade, Investment and Business in Vietnam*, Basingstoke: MacMillan.

VASC (2002), 'Saigon Postel – Don vi thu hai cung cap dich vu dien thoai co dinh' [Saigon Postel – the second provider of fix voice telephone service], online news service, VASC Online, http://www.vnn.vn/i-today/, accessed 28 June 2002.

Vietnam Business Journal – VBJ, Online (http://www.asianmall.com), various issues.

Vietnam Economic Times – VET, Online (http://www.vneconomy.com.vn), various issues.

Vietnam Economy Online 2002, (http://www.vneconomy.com.vn).

Vietnam Posts and Telecommunications News and Magazines, various issues.

Vietnam–US Bilateral Trade Agreement, US Trade Representative, Online (http://www.ustr.gov).

VNPT Website (http://www.vnpt.com.vn).

10. China: Growth and reform in telecommunications

Yanrui Wu

In the past two decades China's telecommunications industry, in common with the rest of its economy, has undergone dramatic changes in terms of the regulatory environment, infrastructure development and provision of services. As a result of these changes, Chinese telecommunications has expanded significantly and China now has the world's second largest number of telephone lines after the United States (ITU 2000). This chapter presents an overview of the recent reforms and the current situation of telecommunications in China. Section 1 reviews the historical growth of the industry; section 2 discusses reforms to the organisational structure; section 3 examines the main players and competition in the market; ownership and privatisation are considered in section 4; the growth outlook for the industry is discussed in section 5; and section 6 presents the chapter's concluding remarks.

1. GROWTH IN CHINA'S TELECOMMUNICATIONS

The establishment of China's telecommunications industry dates from the construction of the first submarine cable by a British company in 1870, and the first landline by the Chinese army in 1879. Services in telegraphs, telephones and wireless communications were then gradually introduced into China, and for decades were dominated by foreign powers (He 1997). Since taking office in 1949, the government of the People's Republic of China has kept tight control on the construction and operation of the telecommunications networks. Growth in the industry during the period 1949–79 was modest. The situation has changed, however, since China's economic reform programme was initiated in 1979, and China's telecommunications industry has grown at a rapid rate during the past two decades. Figure 10.1 shows that the percentage growth in both industry turnover and the number of telephones has been especially robust, and has exceeded the Gross Domestic Product growth rate since the mid-1980s. Industry turnover and telephone number growth rates reached a peak in the early 1990s, but still remained above 30 per cent through to 1999.

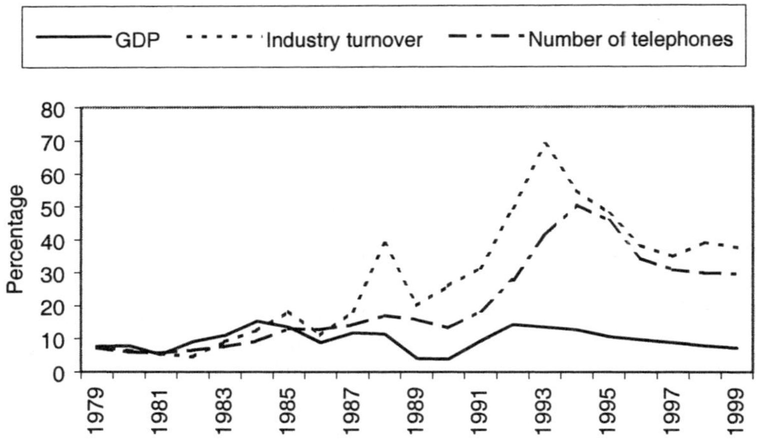

Sources: SSB (1999, 2000)

Figure 10.1 Annual growth in GDP, telecommunication industry revenues and number of telephone services in China, 1979–99

Three factors in particular have contributed to the swift expansion of China's telecommunications: demand; recognition of the value of the sector; and deregulation of the industry. First, telecommunications growth in recent years has been demand driven. Following the implementation of economic reforms in the late 1970s, the Chinese economy achieved an average annual GDP growth rate from 1979 to 1999 of around 9.6 per cent (SSB 2000). This growth has led to a dramatic expansion in economic activity and a consequent substantial increase in the demand for telecommunications services. Since the 1980s, economic and political liberalisation has brought the Chinese people greater freedom to move around the country, and to take jobs away from their hometowns. In particular, millions of rural peasants have found jobs in the cities. In addition, international telecommunications traffic has increased rapidly due to the arrival in China of tourists and offshore investors. These developments have required an expanded telecommunications system to meet the rising demand for both business and personal communications.

Second, since the implementation of the economic reforms, Chinese policy-makers have increasingly recognised from international experience the importance of the telecommunications sector to economic development. Government officials are well informed of international research, for example, the study by Hardy (1980) that shows telecommunications contributes directly to GDP growth. Early in the economic reform process, telecommunications, together with energy and transport, were identified as the main bottlenecks in the economy and were designated as areas for priority development (Yu et al. 1999). A Chinese government report concluded that the marginal contribution to the national economy from investment in information resources was about 16 times greater than that from investment in other sectors (EIU 1997). Traditionally, the role of telecommunications in China was to serve the military and the elites. However, telecommunications came to be seen as an important contributor to national economic growth and, as economic activity increased, was required to provide services to people from all walks of life. These factors resulted in changes in government policy, and by the late 1990s led to an investment boom in the telecommunications sector. Figure 10.2 illustrates the trend of investment in the telecommunications sector during 1990–99. It clearly shows that investment increased dramatically during the 1990s, in particular in 1993, 1994 and 1998.

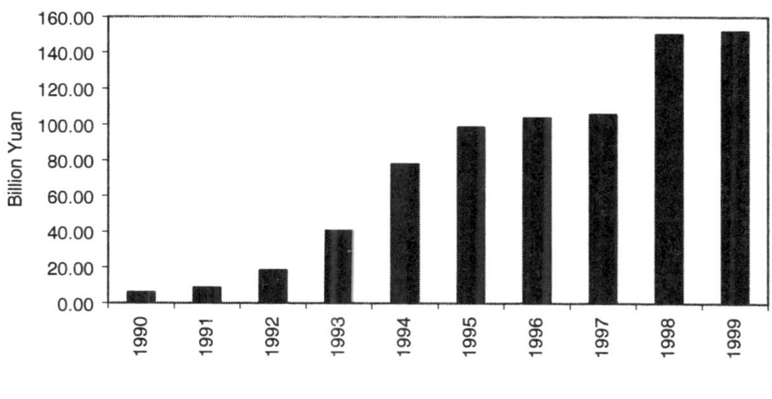

Sources: SSB (2000); Harwitt (1998)

Figure 10.2 Investment in China's telecommunications, 1990–99

Third, recent growth in China's telecommunications has also been driven by deregulation and the emergence of competition. The presence of several domestic carriers and the participation of foreign providers (some of them acting illegally, for example, by providing call-back and internet telephone services) have contributed to the great expansion of the sector as players compete for market share. Competition has also significantly lowered the cost of access to telecommunications services. For example, the installation charge for fixed-line services fell from an average of around 2,571 yuan (around US$320) in 1995, to 500 yuan (US$60) in 1999, and was abolished altogether in 2001. Mobile phone services and access to the internet have also become more affordable. These price reductions dramatically boosted consumer demand, particularly during the 1990s, and have stimulated growth in the sector.

2. ORGANISATIONAL AND STRUCTURAL REFORMS

Accompanying the rapid growth in telecommunications have been reforms to the organisation and structure of the industry. Prior to 1998, several government departments were involved in both operating telecommunications facilities and policy-making in the sector. The Ministry of Posts and Telecommunications (MPT) was the major player and owned, via China Telecommunications Company Limited (China Telecom), the bulk of China's fixed lines. Separate 'private' networks were also owned and operated by other government departments, namely the Ministry of Railways (MOR), Ministry of Electric Power (MEP), Ministry of Defence (MOD) via the People's Liberation Army (PLA), and the Ministry of Petroleum (MOP).

With the implementation of reforms throughout the economy strong pressures arose for the liberalisation and deregulation of telecommunications. An important first step was taken in 1994 with the establishment of a second national carrier, United Telecommunications Corporation (Unicom), which presented a direct challenge to the monopoly held by China Telecom. The major shareholders in Unicom were a number of government departments and public enterprises including MOR, MEP, the Ministry of Electronic Industry (MEI) and the China International Trade and Investment Corporation (CITIC). One of the objectives of the establishment of Unicom was to use the excess capacity of the networks owned by MOR and MEP to provide a range of new services, particularly mobile telephony, paging, internet access and data communications.

However, Unicom experienced major problems in taking on China Telecom, the incumbent carrier and main owner of fixed lines. Although China Telecom was formally separated from MPT in 1994, these two entities remained closely related. This meant that MPT, which held both regulatory and operational responsibility for the industry, failed to act effectively as an independent and transparent regulator. Unicom, as an entity outside MPT, has therefore experienced major problems in interconnection with China Telecom. Size and experience have also been important. Unicom as a new entrant has been too small to compete effectively against its giant and well-established rival.

Restructuring of the central government's administrative system in 1998 involved changes to the organisational structure for telecommunications. MPT, MEI and the State Radio Regulation Commission were merged to form the present Ministry of Information Industry (MII). The aims of this bureaucratic restructuring were clearly to consolidate the government's regulatory role and create an environment for effective competition. As one example, since the restructure, Unicom and China Telecom now fall under the same umbrella of the MII, and may now be able to compete on a more equitable basis. Under the new administrative system, MII is the sole regulator responsible for price setting, licensing and other regulatory matters.

Overhaul of the bureaucratic system has been accompanied by reform of the telecommunications enterprises themselves. One of the major initiatives has been to break up China Telecom's four service divisions into separate entities, with four separately-operated companies now responsible for fixed-line, mobile, paging and satellite communication services. The operator of fixed-line services maintains the name China Telecom (hereafter the two are referred to as 'new China Telecom' and 'old China Telecom'). The mobile service division of old China Telecom was transferred to a new company, China Mobile Communications Group (China Mobile). New China Telecom and China Mobile were both incorporated in May 2000 (*People's Daily* (Overseas Edition) 17 May 2000, 18 May 2000). Another new company, China Satellite Communications Group (China Satellite), was established to provide satellite telecommunications services, while the paging division of old China Telecom was merged with Unicom. Although the operations of the new companies are currently confined to their respective sectors, they will eventually be allowed to diversify into each other's business areas. The rationale for this is to break, in time, the dominant position of old China Telecom, to establish new independent service providers of more comparable size, and hence to promote effective competition in the industry.

3. MAJOR PLAYERS AND COMPETITION

Analysis here focuses on service providers, not equipment manufacturers. The players in China's telecommunications sector comprise the three large corporations already mentioned (new China Telecom, China Mobile and Unicom), as well as a few new smaller firms. In terms of market share, the three big players currently dominate the industry. For the year 2000, the approximate market shares for telecommunications in China were China Telecom 57 per cent, China Mobile 34 per cent, and Unicom 8 per cent

(*People's Daily* (Overseas Edition), 14 February 2001). Other very minor players included Jitong, Great Wall, China Satellite and Netcom. At the end of 2000 two new companies were added to the sector, namely, Shanghai Xintian Telecommunications Limited – which is jointly owned by China Telecom (Shanghai), Shanghai Information Investment Limited, and American Telephone and Telegraph (AT&T) – and Railway Telecommunications Limited (Railcom). The business activities of both new companies will involve value-added services.

The major telecommunications firms in China are set out in Table 10.1. Competition in fixed-line services is limited, as new China Telecom is still the overwhelmingly dominant provider. Unicom is the second largest provider, but, as mentioned earlier, its interconnection with China Telecom has been problematic. As a result, Unicom has been building its own fixed-line network, although its network capacity is still limited. The third owner of fixed lines is the recently established Railcom, which started offering fixed-line services in mid-2001. Railcom relies mainly on the facilities owned by the former Ministry of Railway. These facilities are obsolete and, unless they are upgraded, will constrain Railcom's growth potential. Further deregulation is likely to change the circumstances of fixed-line services. One possibility is the emergence of joint services between Unicom and one or more of the local cable television networks, which currently are not officially permitted to provide telecommunications services. In terms of the length of fixed lines, however, the cable TV networks can already match those of China Telecom. At the end of 1998, the cable TV networks had over 77 million subscribers throughout China, which represented a subscription density greater than that of the telephone services (Duan 2000).

Although the cable TV networks have access to a large pool of households they are suitable mainly for one-way communications only, and are owned and operated by hundreds of separate entities. Another problem is that many networks are not connected with each other. Against government regulations, some small networks have expanded to provide telecommunications services to local communities. For example, the Broadcast Bureau in Hunan created a company that linked 14 of the province's largest cable TV stations into a single network (Lawrence 1999a). The company is now upgrading the new interconnected network to handle two-way transmissions and, eventually, interactive data services.

If regulatory barriers were removed, it is likely that more local cable TV networks would extend their services into the telecommunications sector. For instance, Unicom could upgrade some cable TV networks to provide voice and data communication services. Unicom's target is eventually to have at least a 20 per cent share of the long-distance calls market via its fixed lines. Unicom's expansion in the local call market, however, is very slow, partly because the financial return on local calls is very low. Unicom currently focuses mainly on providing services to large clients such as major companies and government departments, and provides local call services in only three cities. The dominance of China Telecom in fixed-line services is therefore likely to continue for some time.

Table 10.1 Major players in China's telecommunications

Services	Players
Fixed-line	New China Telecom; Unicom; Railcom; cable TV networks
Paging	Unicom; large number of minor providers
Mobile	China Mobile; Unicom; Great Wall; potential entrants (China Satellite, New China Telecom)
Data	New China Telecom; China Netcom; Jitong; Unicom; China Mobile; potential entrants (China Great Wall; China Satellite)
Internet services	Unicom; China Netcom; Jitong; China Mobile; New China Telecom

Note: As at 31 December 2000

China has the world's largest paging service sector. It was deregulated in 1993 and is now the most competitive segment of the nation's telecommunications industry. In 1998 there were more than 2000 providers (Wan 1998), although the number has since fallen due to a wave of mergers and bankruptcies in the sector. In spite of the large number of providers, Unicom has about a 50 per cent share of the market. The direction of further growth is in providing two-way paging (already provided by Unicom in Shanghai) and data communications services.

The mobile services market is dominated by China Mobile and Unicom. China Mobile has a nationwide GSM (Global System for Mobile Communications) network. Unicom also has a GSM network and, in addition, has been licensed to build a nationwide CDMA (Code Division Multiple Access) network. In 1999 Unicom had over 14 cent market share. This rose to 22 per cent by the end of 2000, and is expected to reach 35 per cent by 2005 (Yang 2000). The third provider is China Great Wall Communication Company (Great Wall) which was jointly founded by the former MPT and the PLA. Great Wall intended to build a CDMA network in China but, as the PLA has been banned from business activities, the future of the company is uncertain. It may remain a small player in the market or merge with one of the major mobile carriers.

By the end of June 2001 the total number of mobile phone users in China was about 117 million, more than double the 1999 figure (Table 10.2). China was projected to overtake the United States to become the world's largest mobile phone market by July 2001 (Dwyer 2001). Although the two national carriers have dominated this market, competition between them has already considerably lowered the cost of mobile phone services. Two other licence holders, China Satellite and China Telecom, are also expected to provide services in this area. Future competition in China's mobile phone sector is likely to be intense. In 2001 Unicom, China Mobile and China Telecom were aggressively promoting their wireless internet services under the wireless application protocol (WAP).

Table 10.2 China's telecommunications subscribers (millions)

Year	Telephone	Mobile	Paging	Internet users
1990	6.85	0.02	0.44	–
1991	8.41	0.05	0.87	–
1992	11.47	0.18	2.22	–
1993	17.33	0.64	5.61	–
1994	27.30	1.57	10.33	0.01
1995	40.71	3.63	17.39	0.04
1996	54.95	6.85	25.36	0.16
1997	70.31	13.23	29.69	0.52
1998	87.42	23.86	37.83	2.10
1999	108.81	43.24	47.18	8.90
2000	144.41	85.26	48.64	22.50
2001	164.37	116.76	43.98	n.a.

Note: n.a. = not available

Sources: SSB (2000); CNNIC (2000, 2001); MII (2001)

Data communications are mainly provided by new China Telecom (trading in this market segment as ChinaNet), China Netcom Corporation (CNCNET), Jitong Communications Corporation (ChinaGBN), Unicom (UNINET), and China Mobile (CMNET). Each network has its own comparative advantage in this market. While China Telecom has its existing large customer base, the Unicom network is better positioned in terms of structure and technology. Jitong, established in 1993 by the former MEI and 30 other state-owned enterprises (SOEs) and research institutes, is upgrading its network capabilities. China Netcom Corporation, a new entrant to this market, is providing its services through broadband cable facilities. China Netcom was jointly founded by the Chinese Academy of Sciences, the Ministry of

Railways, the State Administration of Radio, Film and Television (SARFT) and the Shanghai Municipality. According to one report, CNCNET will utilise internet protocol (IP) technology to provide high-speed voice and data services (Lawrence 1999b). With SARFT on its board there is also the possibility that China Netcom will be able to use cable TV lines to link customers with its high-speed network. There are two free education networks providing data communication services, namely the Chinese Education and Research Network (CERNET) operated by the State Education Commission, and the Chinese Science and Technology Network (CSTNET) operated by the Chinese Academy of Sciences. At the end of 2000 data networks for China Great Wall (CGWNET) and China Satellite (CSNET) were under construction. It can be seen from Table 10.3 that, in terms of capacity, ChinaNet is by far the largest of the data communications networks, with a capacity which accounts for about two-thirds of China's total capacity in data communications. The rest, about one-third, is shared among seven other small networks.

Table 10.3 Capacity of China's data communications networks

Trading names	Capacity (megabytes)	Operators
ChinaNet	2 387	New China Telecom
CNCNET	355	China Netcom Corporation
ChinaGBN	151	Jitong
CERNET	117	State Education Commission
UNINET	100	Unicom
CMNET	90	China Mobile
CSTNET	55	Chinese Academy of Sciences
CIETNET	2	Foreign Trade (Government)
CGWNET	under construction	China Great Wall
CSNET	under construction	China Satellite
Total	3 257	

Note: As at 30 June 2001

Source: CNNIC (2001)

There are five internet service providers in China, namely, Unicom, China Netcom, Jitong, China Mobile and new China Telecom. Unicom is the major player with a 50 per cent market share. By the end of 2000 China had about 22 million registered internet users (see Table 10.2). This figure was expected to reach 100 million by 2005 (Weiss 2000). It has been reported that 10 per cent of China's international voice calls were made through IP telephony in

2000, and this proportion was expected to increase significantly to 35 per cent within three years (Masud 2000).

4. OWNERSHIP AND PRIVATISATION

In recent years telecommunications in China has moved away from state ownership and monopoly, to state control through shareholding and competition. The transition in ownership began with the partial floats of China Mobile and Unicom on both domestic and offshore stock markets (see further below). The Chinese government continues to control the two national carriers through various government departments and SOEs. Though value-added services have been deregulated internally, participation of offshore investors and service providers in Chinese telecommunications has been limited by various government regulations. Nevertheless, foreign companies are permitted to sell telecommunications equipment to Chinese corporations and to set up joint ventures for the manufacture of such equipment in China. Large foreign companies such as AT&T, Motorola, Siemens and NEC have taken advantage of this policy, and some of their products, for example, Motorola mobile phones, dominate the Chinese market.

Offshore investors are, however, prohibited from the direct provision of telecommunications services. To circumvent this regulation Unicom devised a unique joint venture scheme called 'Foreign–Chinese–Chinese' (FCC) whereby a separate Sino-foreign joint venture teamed up with Unicom to establish a third, new entity. At one stage a significant proportion of Unicom's equity was accounted for by France Telecom, Sprint, Deutsche Telekom, Bell Canada and NTT. Following government intervention the FCC model was abandoned in mid-1997. Nevertheless, in 1999 some foreign companies were reported to be still providing telecommunications services to Chinese enterprises, for example, by establishing and maintaining Intranet services (Aldrich and Rooth 1999).

Information concerning foreign ownership is limited, although offshore companies are apparently still involved in the businesses of China's main telecommunications service providers. Unicom is reported to have set up a joint venture with the Hong Kong Hutchison Whampoa group, which owns 2 per cent of Unicom. According to the same source China Mobile may team up with the United Kingdom-based Vodafone, and Netcom with Singapore Telecom (Dodgson 2000).

It was mentioned above that China Mobile and Unicom have been partly floated. China Mobile, as a subsidiary of the old China Telecom, offered US$4.2 billion worth of its shares on the New York stock exchange in October 1997, and another US$2.6 billion in 1999. Unicom was listed simultaneously on the New York and Hong Kong stock exchanges in June 2000, raising US$6.4 billion. A total of 20 per cent of Unicom equity is now privately owned. Both new China Telecom and Jitong plan to list their shares on offshore stock exchanges. The current ownership restrictions are likely to be further relaxed in time. In particular, if admitted to the World Trade Organization (WTO), China will be required to permit up to 50 per cent foreign ownership of its telecommunications corporations. Full privatisation of its telecommunications corporations, however, has not yet been placed on China's reform agenda.

5. GROWTH OUTLOOK

Although already one of the largest in the world, the Chinese telecommunications industry seems likely for several reasons to continue its expansion in the coming decades. First, China's level of economic development is still far behind that of the developed economies. The current pace of robust economic growth seems set to continue for at least another decade, and growth-driven demand for telecommunications services is also likely to increase. According to one forecast the number of telephone lines in China will increase from 160 million in 1999, to 370 million in 2010, and to one billion in 2020 (Masud 2000). The number of mobile phone subscribers is expected to reach 250 million by 2004 (Li 2000).

Second, although already very large in aggregate, China's telecommunications sector is still small on a per capita basis. This suggests substantial growth potential. Development in Chinese telecommunications is also highly uneven, both among regions and between urban and rural areas. In 1999 teledensity was 19.2 per cent in the urban areas, compared with only 3.9 per cent in rural China (SSB 2000). The central and western regions also lag behind their coastal counterparts. Table 10.4 presents some summary statistics on the telecommunications sector in the Chinese regions. It shows that China's coastal regions are much better developed in terms of telecommunications than the rest of the county. The table reveals an even greater disparity between rural areas in the coastal regions, and rural areas in the rest of the country.

Third, further deregulation and subsequent increased competition can be expected to continue to expand demand. Pricing in China's telecommunications sector is still controlled by the government, with the prices of services still high by international standards. International call prices are so high that some foreign companies in large cities such as Shanghai and Beijing take the risk of operating illegal call-back services.

Table 10.4 Telecommunications indicators by region, 1998

	Teledensity					GDPpc per capita (US$)
	Urban	Rural	Mobiles	Pagers	Lines	
Coastal						
Beijing	39.3	10.6	9.8	2.4	50.0	2228
Tianjin	29.8	5.5	6.5	6.0	33.3	1698
Hebei	23.7	3.8	2.0	4.9	13.9	783
Liaoning	19.9	6.3	3.6	9.7	19.1	1110
Shanghai	36.1	26.2	9.8	6.1	48.0	3414
Jiangsu	21.8	10.0	2.6	5.6	24.0	1215
Zhejiang	31.3	7.3	3.8	4.7	21.2	1355
Fujian	32.6	6.7	5.3	n.a.	19.8	1224
Shandon	16.9	3.5	2.1	3.4	11.9	983
Guangdong	28.5	6.8	5.1	5.2	19.7	1352
Guangxi	17.9	1.2	1.5	4.7	8.5	567
Mean	27.1	8.0	4.7	4.8	24.5	1448
Central						
Shanxi	16.0	2.4	2.0	3.3	12.5	615
Inner Mongolia	14.2	4.1	2.0	3.3	15.0	617
Jilin	17.4	4.2	2.6	4.9	16.0	718
Heilongjiang	18.4	4.3	3.7	5.1	18.0	910
Anhui	17.4	2.5	1.3	3.3	9.7	556
Jiangxi	18.0	2.6	1.6	4.4	12.2	536
Henan	20.6	3.1	2.1	3.6	13.0	565
Hubei	18.1	3.6	1.7	4.0	12.1	760
Hunan	19.1	2.5	1.6	4.2	9.4	599
Hainan	20.6	1.7	3.2	9.1	10.8	710
Mean	18.0	3.1	2.2	4.5	12.9	659
Western						
Chongqing	20.2	1.8	1.6	3.0	11.2	569
Sichuan	16.4	1.2	1.4	3.3	8.7	512
Guizhou	14.6	0.5	0.8	2.6	6.2	281
Yunnan	29.2	2.8	2.3	7.0	16.7	526
Tibet	24.5	0.1	2.6	3.6	33.9	449
Shaanxi	20.4	1.8	1.5	3.0	11.5	476
Gansu	19.6	1.5	1.3	4.2	14.7	420
Qinghai	18.1	2.5	2.0	3.3	17.9	529
Ningxia	23.3	2.8	1.8	6.6	23.9	515
Xinjing	16.7	8.7	1.8	4.8	24.9	778
Mean	20.3	2.4	1.7	4.1	17.0	505

Note: All indicators except GDP per capita represent teledensity per 100 persons

Source: SSB (1999)

China Telecom recently changed its practice of charging for local and long-distance calls by shortening the length of the charging 'blocks' to one minute for local calls, and to six seconds for distance calls. (For example, whereas previously a local call lasting three minutes and one second would be charged as six minutes, it is now charged as four minutes.) Mobile phone calls, however, are still double-charged, that is, to both the person making the call and the receiver of the call. It is likely that the price of telecommunications services will continue to fall as the market becomes more competitive.

Notwithstanding the factors discussed above, continued growth in Chinese telecommunications remains subject to a number of constraints and uncertainties:

- The Telecommunications Act, promulgated in September 2000, is the first Chinese legislation to specify a complete set of regulations aimed at promoting efficient allocation of resources and fair competition, and at ensuring universal service obligations in any industry. It remains to be seen how effectively this legislation will be enforced.
- There is concern relating to the MII's joint role of telecommunications regulator and majority shareholder in a number of key industry players. This creates an obvious conflict of interest that has the potential to adversely affect the independence and transparency of its regulatory role.
- Regulations on foreign investment have recently been characterised by frequent changes and ambiguities, with the resulting confusion causing some foreign participants to reconsider their plans for the Chinese market.
- There are considerable problems with new China Telecom, the dominant fixed-line operator. It is heavily in debt and many of its local branches are unprofitable due to poor management and overstaffing. A significant improvement in its performance is crucial for infrastructure development and the continued growth of the industry.

6. CONCLUDING REMARKS

The dramatic growth experienced by the telecommunications industry in China throughout the 1990s looks set to continue. This growth has been driven largely by deregulation of the industry and increased consumer demand. Chinese telecommunications is dominated by three main providers, and there is some competition in the industry. Some areas such as paging, mobile phones and data communications are more deregulated and more competitive than others. Deregulation of the industry, however, is far from complete. For a long time the industry has been directed mainly by fragmented administrative decrees, and telecommunications legislation is still in its infancy. The first Telecommunications Act was introduced only in September 2000, and its enforceability is yet to be tested.

Many fundamental issues such as property rights, foreign participation and interconnection remain unresolved. Some issues such as private ownership are associated with the transitional nature of the Chinese economy. Others such as interconnection are specific to the industry. At present competition is largely confined to wireless and other value-added services, with little competition in the fixed-line sector. Interconnection was not effectively enforced at the early stages of development, and has become a bottleneck problem for the industry that may take some time to be resolved.

It is anticipated that recent reforms to the Chinese economy, and China's imminent membership of the WTO, will lead to further deregulation and growth in the telecommunications sector. Indeed, the various indicators examined in this chapter point to a bright growth outlook for the industry. Given the significance of the industry for the development of the national economy and of China in the global economy, it can be expected that further growth of telecommunications will boost China's overall economic growth and create vast opportunities for international trade.

ABBREVIATIONS

AT&T	American Telephone and Telegraph
CDMA	code division multiple access
CERNET	Chinese Education and Research Network
CGWNET	China Great Wall Network
CITIC	China International Trade and Investment Corporation
CMNET	China Mobile Network
CNCNET	China Netcom Corporation
CNNIC	China National Network Information Centre
CSNET	China Satellite Network
CSTNET	Chinese Science and Technology Network
EIU	Economist Intelligence Unit
FCC	Foreign–Chinese–Chinese joint venture
GSM	global system for mobile communications
IP	internet protocol
ITU	International Telecommunications Union
MEI	Ministry of Electronic Industry
MEP	Ministry of Electric Power
MII	Ministry of Information Industry
MOD	Ministry of Defence
MOP	Ministry of Petroleum

MOR	Ministry of Railways
MPT	Ministry of Posts and Telecommunications
PLA	People's Liberation Army
SARFT	State Administration of Radio, Film and Television
SOEs	state-owned enterprises
SSB	State Statistical Bureau
UNINET	Unicom Network
WAP	wireless application protocol
WTO	World Trade Organization

REFERENCES

Aldrich, M. and A. Rooth (1999), 'Market opportunity: an investment threat', *Telecommunications*, **33** (10), 168–70.

CNNIC (2000), Internet survey results at the end of December 2000 (http://www.cnnic.com.cn).

CNNIC (2001), Internet survey results at the end of June 2001 (http://www.cnnic.com.cn).

Dodgson, C. (2000), 'China looking for the great leap forward', *Communications International*, **27** (5), 37–8.

Duan, N. (2000), 'Trend of national telecommunications network development', Telecommunications Research Institute, MII, mimeo.

Dwyer, Michael (2001), 'Wireless warriors zero in on China', *Australian Financial Review*, 20 July, 24.

EIU (1997), 'How has China financed its telecoms build-out?', *Business China*, 24 November, 7–9.

Hardy, A.P. (1980), 'The role of the telephone in economic development', *Telecommunications Policy*, **4** (4), 668–80.

Harwitt, E. (1998), 'China's telecommunications industry: development patterns and policies', *Pacific Affairs*, **71** (2), 175–93.

He, Z. (1997), 'A history of telecommunications in China: development and policy implications', in P.S.N. Lee (ed.), *Telecommunications and Development in China*, Cresskill, NJ: Hampton Press, pp. 55–88.

ITU (2000), *International Telecommunications Indicators*, Geneva: ITU.

Lawrence, S.V. (1999a), 'Telecommunications brawl', *Far Eastern Economic Review*, 30 September, 69–71.

Lawrence, S.V. (1999b), 'China Netcom Corp. breaks the mould', *Far Eastern Economic Review*, 30 September, 70–1.

Li, P. (2000), 'Wireless internet connection in China', *Sanliangshenghui Weekly*, **11**, 46–7.

Masud, S. (2000), 'In China: opportunities abound, but perils remain for equipment vendors', *Telecommunications*, **34** (1), 9–10.

MII (2001), 'Telecommunications update: June 2001', Ministry of Information Industry (http://www.mii.gov.cn/hyzw/tongji/yb/tj0602.html).

SSB (1999), *Comprehensive Statistical Data and Materials on 50 Years of New China*, Beijing: Statistical Publishing House.

SSB (2000), *China Statistical Abstract*, Beijing: Statistical Publishing House.

Wan, X. (1998), *Information Age: China's Choices in the 21st Century*, Beijing: Social Sciences Documentation Publishing House.

Weiss, E. (2000), 'Internet growth hastens network upgrades', *Telecommunications*, **34** (5), pp. 22, 28.

Yang, T. (2000), 'Unicom listed in New York and Hong Kong stock exchanges', *China Business Times*, 23 June, 1.

Yu, Y., B. Zheng and S. Hong (eds) (1999), *The Research Report on China's Entry into WTO: The Chinese Industries*, Beijing: Social Sciences Documentation Publishing House.

11. India–Singapore synergy in telecommunications

Faizal Yahya

1. INTRODUCTION

In recent years, the Asia–Pacific region has attracted a lot of attention from the world's major telecommunications players because of the reforms taking place there and the opportunities thus created for foreign investment. As Asian countries liberalise their telecommunications sector, a major challenge for them is to ensure that the opening up of this sector will bring about greater benefits than adverse consequences. In particular, many countries are concerned about the prospect of the domestic market coming under increasing dominance by foreign interests.

In response to this concern, India's strategy is control through legislative acts, policy regulations, joint ventures and restrictions on the amount of shares a foreign investor can own in a local telecommunications company. At the same time, the need to modernise the sector has prompted the Indian government to make efforts to entice foreign investments and collaboration on acceptable terms.

In the meantime, the liberalisation of the telecommunications sector in Singapore has spurred the dominant local player, Singapore Telecommunications Private Limited (SingTel), to look beyond Singapore's national boundaries and invest overseas. Thus considerable potential synergy exists between Singapore and India in this area. Indeed, SingTel has entered the Indian market through a Joint Venture (JV) with the Bharti Corporation and has invested substantial amounts of funds in the JV. These recent experiences suggest that there exists great scope for Asia–Pacific economies to collaborate and exploit synergies in telecommunications for mutual benefit.

The remainder of this chapter is organised as follows. Section 2 presents some background information, before a brief review of recent developments in the Indian telecommunications sector is undertaken in section 3. A similar review of developments in Singapore is presented in section 4. In Section 5 we examine the India–Singapore experience with collaboration in this sector, and section 6 concludes the chapter.

2. BACKGROUND

The World Trade Organization (WTO) estimated that, in 1995, global telecommunications revenues amounted to some US$601.9 billion, representing 2.1 per cent of world gross domestic product (GDP). In the era of globalisation, an increasing number of countries have turned towards market-oriented approaches to liberalise this important sector. Underlying this trend has been a number of key factors, including the inability of state-run companies to keep abreast of accelerating technological change, a growing political consensus on the benefits of liberalisation, and increasing pressure on government budgets.

Among developing countries, the tasks of modernising and expanding telecommunications networks are made more challenging by a lack of capital for investment in infrastructure projects. The problem is compounded by the fact that in some countries such as India, varying degrees of access to the telephone have become indicative of a growing disparity between the wealthy and the poor.

Among international telecommunications companies, the growing challenge is the competition to enter the potentially lucrative Asian market, where telephone density is only about one-half of the world's average and where there is a huge population base, growing wealth, and momentum in economic liberalisation. For example, while demand for mobile phones is levelling off in North America and Europe, the number of mobile users in the Asian region is expected to continue rising rapidly in the next few years.

Of course, the United States, the European Union, and Japan remain the world's largest telecommunications markets: as shown in Table 11.1, together they accounted for almost three-quarters of global revenues, 65 per cent of global investment, and 63 per cent of international traffic in 1995. Nevertheless, the fast expanding markets in Asia ensure that Asian telecommunications companies are among the most profitable in the world. As Table 11.2 illustrates, of the top 10 international telecommunications companies in terms of revenue earned, only two are Asian companies, NTT and NTT DoCoMo (Divyanathan 2002). However, as shown in Table 11.3, of the 10 most profitable companies, more than half are Asian firms.

By now there is already a sizeable literature on recent developments in the telecommunications sector of India. For example, Singh (1999) reviewed the colonial beginnings of Indian telecommunications while Chowdary (1997) focused on the period immediately after independence in 1947. The post-independence period was characterised by heavy government intervention which tended to impede the development of the sector. Vested interests, in the form of the bureaucracy as well as the privatised incumbent companies, also tended to stifle competition and expansion. Mani (2000) described the deregulation process in India's economy generally during the 1990s and how it has impacted on the telecommunications sector in particular. By and large the general economic reforms and liberalisation process which began in 1991 resulted in an accelerated privatisation of the telecommunications sector. Vittal (1995) investigated this process of privatisation, while Singhal (1997) examined the marketing of the Indian telecommunications sector to private and foreign investors.

Table 11.1 Share of telecommunications revenue for the EU, Japan and the United States (%)

	European Union	United States	Japan	Total
Revenue	28.3	29.7	15.6	73.6
Main lines	26.1	23.8	8.8	58.7
Investment	27.1	15.6	22.0	64.6
International traffic	35.2	25.3	2.0	63.2

Source: World Trade Organization (1997)

Table 11.2 Top ten telecommunications companies – ranking by revenue

Ranking	Company	Revenue (US$billion)
1.	NTT (Japan)	93.4
2.	Verizon Communications (USA)	67.1
3.	SBC Communications (USA)	54.3
4.	AT&T (USA)	52.6
5.	Deutsche Telekom (Germany)	43.3
6.	NTT DoCoMo (Japan)	41.4
7.	France Telecom (France)	38.5
8.	WorldCom (USA)	35.2
9.	Vodafone Group (UK)	32.7
10.	Telefonica (Spain)	27.8

Source: Straits Times, Singapore, 16 October 2002

Much has also been written about Singapore's experience with telecommunications liberalisation. For example, Heng and Low (1990) argued that Singapore needed to increase competition in the sector in order to accommodate rapid advances in technology, ideological shifts and the challenges of globalisation. Chen and Kuo (1995) examined the importance of the sector as a critical component of economic development. Mansell and Jenkins (1992) documented the successful launch of Tradenet in January 1989. This electronic network made use of Singapore's good telecommunications infrastructure to process electronically over 90 per cent of Singapore's trade and customs regulations. Joo (1993) examined initiatives undertaken to project Singapore as a telecommunications hub and its overseas ventures as a key regional player. Mesher and Zajac (1997) reviewed the modernisation of the Singapore telecommunications sector through the provision of services such as the internet and cellular telephony.

Table 11.3 Top ten telecommunications companies – ranking by profit margin

Ranking	Company	Profit margin (US$billion)
1.	Taiwan Cellular Corp	37.07
2.	Swisscom	35.02
3.	Emirates Telecommunications Corporation	31.84
4.	China Mobile	28.65
5.	PT Telkom	26.35
6.	SingTel	22.14
7.	Telmex	21.17
8.	Chunghwa Telecom	20.46
9.	VSNL	19.79
10.	Hutchison Whampoa	19.67

Source: Straits Times, Singapore, 16 October 2002

This chapter will differ from the above works, in that it seeks to examine the linkages between India and Singapore in the telecommunications sector. In particular, it will examine the motivating factors behind India's quest to attract foreign investments as well as Singapore's need to regionalise its economic activities, and to create an 'external economic wing' (Huat 1995). In this regard, the analysis will follow the lines established by Asher and Rajan (1994) who provided a seminal analysis of Singapore-India economic relations, but in this case there will be a particular reference to telecommunications.

It may be useful at this stage to compare the size and potential of the relevant markets in India and Singapore. As Table 11.4 shows, India's potential market dwarfs that of Singapore. Indeed, if the Indian government's ambition of raising the teledensity of India to seven per 100 people materialises, the country could rank among the biggest seven or eight telecom markets in the world (*Business Times*, 8 August 2000). At the same time, as a survey conducted by the Telecommunications Regulatory Authority of India (TRAI) indicates, the overall quality of basic telecommunications services in India remains poor (*Economic Times*, 24 August 2002). In contrast, the telecommunications sector in Singapore is regarded as one of the best in the Asia–Pacific region. These contrasts offer considerable scope for complementarity and synergy.

Table 11.4 Comparison between India and Singapore telecommunications markets, 2001

	Population	Per capita GDP (US$)	Telephone density (per 100)	Cellular density (per 100)
India	1.02 billion	435	3.2	0.19
Singapore	4.3 million	26,500	47.2	69.2

Source: Office of Telecommunications, International Trade Administration, US Department of Commerce, Government of the United States of America

3. DEVELOPMENTS IN INDIA

For India, the advent of telecommunications offers the possibility of geographically distributing its economic activities more evenly across the nation rather than being constrained by the traditional predominance of industrial states such as Maharashtra and Gujarat. In recent times, it has also allowed the more advanced economic sectors, particularly information technology (IT), to play a significant role in global markets despite the handicap imposed by the country's general inadequacies in terms of basic transport infrastructure such as roads, railways, air transport and seaport facilities.

3.1 Drivers of Reform

India's venture into the realm of telecommunications began with the India Telegraph Act of 1885. Despite this early start, the telephone service in India nearly a century later (around the beginning of the 1980s) was recognised as one of the worst in the world. The telephone was considered a luxury and the teledensity ratio was only about 0.5 telephones per 100. Telecommunication facilities were generally concentrated in urban areas. Mumbai, for example, had the best telephone and data access capabilities. Major cities such as New Delhi and Bangalore were also fairly well served. By contrast, services in remote rural areas were few and far between. The major hindrances to the development of the sector were mainly regulatory in nature. While the infrastructure needed overall improvement and upgrading, it received scarce attention from the bureaucrats. The government collected substantial revenues from the profits of mammoth state-owned telecommunications enterprises and showed little inclination to strive for maximum efficiency (Slater 2002).

This attitude began to change in the late 1980s with a dawning realisation that telecommunications was a pivotal sector that could play a critical role in economic development. Under the late Prime Minister Rajiv Gandhi, the government recognised that in order to develop the country's software services sector, it was necessary to rely on the telecommunications sector as a basic infrastructure. Although India's entry into the global software business predated the telecommunications deregulations of the 1990s, the liberalisation of the sector has evidently assisted the continuing expansion of the country's presence on international IT software markets. In turn, the spectacular growth of the software industry has continued to fuel a huge demand for a range of high-end telecommunications services, especially in fast-growing metropolitan areas such as Delhi, Karnataka and Andhra Pradesh. More generally, liberalisations in the telecommunications sector became part of a broader wave of economic reforms sweeping through India, as well as a number of other developing countries, during the last decade of the twentieth century.

3.2 Reforms During the 1990s

During this decade, the government issued two sets of National Telecommunications Policies (NTP) aimed at expanding and improving services, by establishing guidelines for the entry of private sector interests (both domestic and foreign) and providing universal access.

The objectives of the 1994 NTP were to ensure the availability of telephones on demand, universal coverage of all villages and the provision of world-standard telecommunications services. The goal was to enable the country to emerge as a major manufacturing base and a significant exporter of telecommunications equipment, while at the same time safeguarding the defence and security interests of India.

By the late-1990s, it had become clear that the liberalisation policies foreshadowed in the 1994 NTP had not been fully implemented according to plan. Although there had been a rapid expansion of the mobile phone networks in the cities and urban areas, most of the projects were facing difficulties. In some cases, insufficient returns on investments meant that the full operational costs of the projects could not be recovered. In addition, basic services had not expanded significantly in rural and remote areas. To promote more rapid development, the government decided to issue 1999 NTP to replace 1994 NTP.

The targets of 1999 NTP included: making fixed telephone connections on demand by the year 2002 and sustaining it thereafter so as to achieve a teledensity of seven by the year 2005 and 15 by the year 2010, encouraging the development of telecommunication in the rural areas and making it more affordable through a suitable tariff structure, and making rural services mandatory for all fixed service providers. A key aim was to increase rural teledensity from the then prevailing level of 0.4 to 4 by the year 2010 and to provide reliable transmission media in all rural areas. Another goal was to provide internet access to all district headquarters by the year 2000 and high-speed data and multimedia capability using technologies including ISDN to all towns with a population greater than 200 000 by the year 2002.

While the NTPs have not been fully successful, in that some targets have not been fulfilled, the liberalisations to date have produced a number of positive outcomes, including an expansion and upgrade of services, and a substantial lowering of prices.

The establishment of the TRAI in 1997 as an independent regulator was a key step towards the effective implementation of reforms in the sector. From that point on, there was a separation between the role of the government as the regulating authority, and its role as the dominant service provider. This separation reduced the obvious potential for conflicts of interests to arise.

It was also in line with efforts to demonopolise the industry and to allow the participation of private interests. The government monopoly on basic plain old telephone services (POTS) was lifted in 1995. At the same time, new licensing procedures for private operators were established. The mobile phone sector was opened up to private companies in 1996, followed by the liberalisation of national long-distance operations in 1999. Following that date, there were no longer any restrictions on the number of players in this segment of the market. In the same year, internet service provider (ISP) companies were allowed to tap into international telecommunications cables.

In preparation for these moves, the government's telecommunications operators had been corporatised and the resulting corporations soon became natural candidates for partial privatisation. Mahanagar Telephone Nigam Limited (MTNL) was established in the late 1980s with the head office being based in Mumbai, India's main commercial and financial city. Videsh Shanchar Nigam Limited (VSNL) came into existence in 1986 (its head office was also located in Mumbai) and was licensed to be the sole provider of international services (Sridhar 1998). VSNL offered a variety of services including telephony, telex, telegraph, internet access, packet switched data transmission, video conferencing and others, but its main business is telephony.

The Department of Telecommunications Services (DTS) was second only to Indian Railways in terms of staff strength. As of the year 2000, it employed 400 000 workers (*Business Times*, 2 October 2000). Despite considerable opposition from many of these workers and their trade unions, the Department was transformed into a government-owned corporation, Bharat Shanchar Nigam Limited (BSNL). The affected workers were provided assurances regarding their pension benefits and job security. BSNL's new status is expected to help it reduce bureaucratic red tape, increase efficiency and gain greater access to capital for expansion and upgrading of capacity.

Communications Minister Ram Vilas Paswan compared such transformation to 'a rebirth in Hindu tradition'. He added that, 'the more the competition, the better it is for the consumer' (*Business Times*, 2 October 2000). These remarks underline the expectations of BSNL, and of the government, that there would be more rapid growth, lower tariffs, and improved services.

It has been suggested that BSNL is overstaffed by a factor of at least three, relative to its 24 million-phone network. On the positive side, however, its assets value of Rs680 billion as well as its annual revenue of US$3 billion are quite attractive to international investors. BSNL's debt equity ratio of 1 : 1 would also be a plus in fund raising (*Business Times*, 2 October 2000).

As stated above, VSNL was granted a sole licence to provide international services. This status was to remain until at least March 2003. However, the WTO agreement on Telecommunications Services in February 1997 changed the scenario completely. As part of the agreement, India committed itself to liberalise its market in phases. In April 2002, the Indian government scrapped VSNL's monopoly on international telephony and allowed a private company, Bharti Tele-Ventures, to enter this market segment (Slater 2002).

VSNL is 53 per cent government owned and is one of India's most cash-rich firms, with nearly US$1.06 billion in reserves. When fully privatised, VSNL stands to gain from more financial reserves, a ready-made infrastructure and an overseas listing on the stock market. The government has also promised VSNL a free licence to enter the national long-distance telephony business and permission to expand its internet business across India as compensation for ending its monopoly over overseas telephone calls. With just over 3000 employees, VSNL does not have the problem of a bloated staff structure like most state-run companies.

3.3 Mobile Telephone Networks

India is placed among the top 10 countries in the Asian region in terms of the size of its mobile phone market. As of May 2003, India had 14.17 million mobile phone users. This is a huge increase from 2.5 million subscribers in August 2000 and 4.54 million subscribers in August 2001. Such rapid growth can be explained primarily by the deregulation of the mobile phone market segment and consequent reductions in cellular tariff rates. As of September 2001, the peak rate for cellular charges was Rs4.5 per minute, compared with Rs16.8 per minute in May 1999.

While this trend is encouraging, there are conflicts emerging between the cellular operators and the DOT. The main problem is that the revenue collected by the DOT from the large licence fees paid by cellular operators was not made available to the DOT for network deployment. A number of these mobile operators have indicated that their positions were untenable within the existing structure and requested the government to defer the due date for their periodical licence fee payments. As a response, the government announced in December 1997 that mobile operators could assign their licences to financiers as security. It also raised the external borrowing limit for operators from 35 per cent to 50 per cent of the projected costs.

The DOT classified the Indian market according to subscriber potential. Thus, the national market has been divided into 'metro', 'A', 'B' and 'C' circles. According to the Cellular Operators Association of India (COAI), the 'C' category expanded by 155 per cent during the year from January 2000 to January 2001, almost twice as fast as the more lucrative 'metro' circles, which grew by 75 per cent in the same period. This indicates that the lower end of the market has been rapidly catching up with the higher end. As cellular services become more affordable, the government plans to allow two more companies to enter each telecommunications circle. One company will be government owned (either MTNL or BSNL) while the other will be from the private sector (*The Hindu* 6 January 2001).

The government decided to auction a new licence for a fourth cellular operator in 2001 after a gap of five years. The bidders must be Indian companies with a foreign equity ceiling fixed at 49 per cent and the successful bidder would have to pay 17 per cent of its earnings as licence fee in addition to the one-time entry fee of Rs10 million (US$200,000) (*Business Times*, 3 May 2000). The fourth cellular operator would be allotted frequency in the 1710–1785 Megahertz (Mhz) slot paired with the 1805–1880 Mhz slot. In addition, the government also invited bids for three vacant slots in the Andaman & Nicobar Islands and one in West Bengal. The licences would be issued for 20 years. The COAI supported the government's move, stating that it expects this to lower charges while improving service, through greater competition.

3.4 Bharti Enterprises

Bharti Enterprises Limited provides a case study of Indian private companies in the telecommunications industry. It is rapidly becoming a major player in India's telecommunications sector. Its founder, chairman and managing director is Sunil B. Mittal. In 1985, Bharti Enterprises became involved in the telecommunications business by manufacturing push-button telephone sets. A turning point for Bharti was 1994 when the company acquired the cellular licence for New Delhi. At the time, the company was able to assemble a consortium comprising France's Vivendi and Mauritius-based Emtel to win the licence despite strong competition.

Starting with the networks in Delhi and the northern state of Himachal Pradesh, Bharti began acquiring money-losing cellular networks in the states of Karnataka, Tamil Nadu, and Andhra Pradesh and managed to make a profit from running them. Bharti also established India's first private fixed-line network in the state of Madhya Pradesh. It has been engaged in talks to buy stakes held by Luxembourg's Millicom and BellSouth in Skycell Communications, in a bid to acquire complete control of the Madras-based mobile operator. Bharti was the first Indian telecommunications group to venture overseas with a foray in the Seychelles.

Mittal has formed a holding company called Bharti Telecom to manage his telecommunication's portfolio. Bharti Telecom has two majority owned subsidiaries, Bharti Televentures and Bharti Telespatial. In turn, Bharti Televentures is the holding company for three cellular operations: Bharti Cellular operates in Delhi, Bharti Mobile operates in Andhra Pradesh and Karnataka, and Bharti Telenet operates a cellular service in the 'C' circle of Himachal Pradesh. As of June 2000, Bharti had more than 800 000 customers and 4000 employees.

In August 2001, Bharti Enterprises won eight of the 17 new mobile licences issued following the July 2001 government auction. As a result, Bharti has extended its coverage to Mumbai, Maharashtra, Gujarat, Haryana, Western Uttar Pradesh, Madhya Pradesh, Kerala and Tamil Nadu.

To expand its urban network, Bharti has added Spice Cell and its business in Calcutta to the growing array of licences that it has acquired from government auction. Bharti purchased Spice Cell in an all-cash deal that cost about US$100 million. By acquiring Spice Cell, Bharti has secured a foothold in India's lucrative metropolitan markets. Prior to this, Spice Cell was 51 per cent owned by ModiCorp, a holding company of India's B.K. Modi business family and 49 per cent owned by Hong Kong's Distacom.

Bharti is listed on the Indian stock exchange. As for the shareholding pattern of Bharti Televentures, a 54 per cent stake in the company is held by Bharti Telecom, 18 per cent equity is owned by Singapore Telecommunications and 21 per cent is owned by Warburg Pincus (one of the world's largest equity investors), with the balance being owned by other, small stakeholders (*Economic Times*, 10 September 2001).

4. DEVELOPMENTS IN SINGAPORE

In the late 1980s, the Singapore government developed an overall strategy to liberalise the island state's domestic economy and to encourage its homegrown companies to venture into overseas markets, and the telecommunications sector was included as part of this overall strategy. Singapore has a relatively small domestic market and lacks a natural market hinterland. It has no alternative but to create an external economic wing for itself.

At this time the government initiated a strategy to double the country's IT and telecommunications revenues to US$232 million by 2005. Its most immediate concern was to transform the incumbent operator, Singapore Telecommunications from a protected, government-owned monopoly into a nimble pan-Asian operator. As a result, SingTel transformed into an incorporated body in March 1992.

In 1995, SingTel implemented its initial public offering (IPO) and raised over S$4 billion. Since then, the government through its investment arm, Temasek Holdings, has retained a majority interest in SingTel. As of March 2003, Temasek Holdings held a 67 per cent stake in the corporation (*Business Times*, 8 March 2003). SingTel has been assisted by increasing investor confidence. For example, in November 2001, its bonds went on sale and were expected to raise S$1.5 but investor interest was such that S$2.3 billion was raised instead (Shari 2001).

In 1997, the government started to open up the telecommunications sector by allowing the entry of the private, foreign-owned company M1 into the market. Subsequently, the domestic market was made even tougher for SingTel, when the government decided to bring forward the introduction of full market competition from 1 April 2002 to 1 April 2000 (Ministry of Communications and Information 2000). The government also indicated that it would continue to liberalise the market after this event and would welcome foreign investment, to ensure effective and sustainable competition.

A major motivating factor in support of the accelerated deregulation of the Singapore market was the need to fulfil sectoral liberalisation commitments with the WTO. This was followed by bilateral negotiations for the Free Trade

Agreements (FTAs) between Singapore and various foreign countries. Singapore needed to liberalise services sectors as it had relatively few tariffs on manufacturing goods to attract potential FTA partners.

In 2000, WTO began a new round of discussion with member countries to liberalise their telecommunications markets. Singapore had to deliver on its commitments to open up its domestic market to foreign investors and collaboration. As scheduled, the government opened up the telecom sector and private companies such as StarHub, MobileOne, Yuana and Flag Telecom were awarded licences to establish rival telecommunications operations.

To meet the challenges of further innovations in the global telecommunications market, telecommunications services are constantly undergoing improvement. Three statutory boards – including International Enterprises (IE) which was previously known as the Trade and Development Board (TDB), the Economic and Development Board (EDB) and the Infocomm Development Authority (IDA) – were given the task to collaborate and expand the use of broadband services. Their role was to foster innovation and to lead in developing e-commerce with the ultimate aim of establishing their departmental expertise in this area. This would enable Singapore to profile itself as one with a track record in e-services and to bid for projects in the emerging markets of China and India that are poised to open up.

According to the Deputy Prime Minister, Dr Tony Tan, Singapore aims to develop itself as a global hub for business and investments (Yam 2001). A major part of this strategy involves enhancing Singapore's ability to compete in the rapidly liberalising global telecommunications sector.

In its efforts to nurture home grown multinational companies (MNCs), the government targeted SingTel as one of the best placed with the potential to make the transition from operating in tiny Singapore into a Pan-Asian corporate giant.

As early as 1994, SingTel's president and chief executive officer (CEO) Lee Hsien Yang had noted that the company had negligible foreign investments and begun planning for it to expand and diversify its revenue sources beyond Singapore (Yang 2002). This was in line with the expectation that the opening up of the domestic telecommunications market and the entry of foreign players would result in a squeezing of SingTel's profit margins. Moreover, it was recognised that unless the company grew beyond its home base, it would not be big enough to compete with major international players (Hansen and Shameen 2001).

Fundamentally, SingTel had to transform itself from a statutory board with a monopoly position in the local business, into a publicly listed and commercially motivated organisation capable of competing on an international level. One of the key challenges during its transformation was the introduction of time-based charging for local calls in December 1991 (*Straits Times*, 27 August 2001). Previously, SingTel had implemented a flat rate, and the new regime resulted in a public outcry initially but it was successfully implemented nevertheless. Since then, SingTel has helped to turn the island state into one of the most wired countries in the world.

The regionalisation strategy developed by SingTel mirrored the transformation of the Asian telecommunications market in the 1990s and the changing trends of the global market. When SingTel first started its overseas expansion, it adopted an opportunistic strategy and forayed beyond the Asian region because most of the Asian market was closed and tightly regulated. In addition, the company began investing in smaller investment projects and gradually increased the size of its investments as it began to accumulate overseas experience. When SingTel invested over S$900 million in the Belgian carrier Belgacom as one of its earlier European investments, this was at the time the largest investment made overseas by a Singapore company.

SingTel also invested in other European countries, such as the UK and Norway, but gradually sold it stakes in them to concentrate on business opportunities in the Asian region, this provided more scope for synergy with joint venture partners and where SingTel could more readily develop a strong and recognisable brand name. For example, in 1999, SingTel sold its Norwegian investment Netcom to TeleDanmark for S$550 million.

A series of changes was also implemented to capture a larger portion of corporate customers' regional telecommunications traffic. For example, in early 2001, SingTel introduced an internet protocol (IP) virtual private network (VPN) service as part of a S$40 million (US$23.8 million) package of enhancements to the ConnectPlus range of products. This range of products included internationally leased circuit services, frame relay, corporate voice and router management in 11 neighbouring countries, with direct links to the United States and the United Kingdom. SingTel also planned to double the size of its S$100 million (US$59.5 million) investment in corporate network infrastructure by the end of 2001. In 2000, the frame relay and leased circuit services accounted for approximately 80 per cent of the company's US$378 million revenue from data services. By the end of 2002, SingTel estimated that frame relay, leased circuits and IP would each account for one-third of its total data services revenues. SingTel has managed to offset its losses in the international calls services sector by growth in the data and wireless business.

SingTel plans to boost its offshore earnings from 27 per cent to about 50 per cent of total earnings. It has accumulated investments approximating S$5 billion in more than 80 joint ventures in 20 countries and territories, including China, India, Malaysia, Indonesia, the Philippines, Thailand, Vietnam, Taiwan, Australia, the United Kingdom and the United States. The company has grown at a remarkable rate, from barely 200 000 cellular phone subscribers to over 20 million customers in the Southeast Asian region alone. However, SingTel has encountered its share of disappointments as part of its expansion history.

For example, its foray into the Hong Kong market to bid for Cable & Wireless HKT was thwarted by Pacific Century Cyberworks (PCCW) of Hong Kong in March 2000. SingTel's bid failed because Chinese banks and companies backed PCCW's rival bid (Saywell 2001b). SingTel also made two unsuccessful attempts to move into the Malaysian market: it lost out to British Telecom in a bid for Maxis in 1998 and then lost again in its bid for a stake in Time dotCom in 1999. Time dotCom was the political arm of Malaysia's politically connected Renong Group. Given that SingTel at the time was 78 per cent owned by the Singapore government through Temasek Holdings, it became a national security issue for Malaysia and this resulted in a decision by then Malaysian Prime Minister Mahathir Mohamad to oppose the proposed deal (Hansen and Shameen 2001).

Unlike its failure to acquire Cable & Wireless HKT and Time dotCom, SingTel managed to secure C&W Optus (Australia) in a US$7.5 billion deal. This acquisition has gone a long way toward transforming SingTel into a pan-Asian telecommunications group, with interests ranging from India to Australia. The follow through to the Optus deal is likely to be an increasing interest by the company in the Indian telecommunications market.

5. INDIA–SINGAPORE COLLABORATION

At an inter-governmental level, Singapore and India formed a joint task force in the area of communications and information technology (ICT) in March 2000. The task force was chaired by the Permanent Secretary of Ministry of Communications and Information Technology (MCIT) of Singapore and the Secretary of the Ministry of Information and Technology of India. The second meeting of the task force was held in July 2001 in Singapore. It outlined a number of measures to foster closer ICT collaboration between India and Singapore.

The two governments have continued their official exchanges and encouraged their respective private sector to strengthen ICT linkages with each other. To promote closer linkages in the area of telecommunications, the IDA and the Confederation of Indian Industries (CII) signed a memorandum of understanding (MOU) in November 2000 in New Delhi. The MOU was in line with IDA's strategy of fostering strategic partnerships and alliances with overseas industry partners. In April 2002, the Singapore and Indian governments signed two MOUs in the field of telecommunications and culture. The purpose of the telecommunications MOU was to facilitate the exchange of information in areas of policy and regulation, technologies, current international legislation related to telecommunications, spectrum management and standardisation.

Some of these official intentions have been turned into commercial realities. In August 2000, SingTel spent US$400 million for an effective 30 per cent share in the telecom stake of Bharti Enterprises. This investment translated into ownership of 20 per cent of Bharti Telecom Limited and 30 per cent of Bharti Televentures.

In October 2000, the IDA led a delegation of fifteen Singapore Infocomm firms and three government agencies in attending the IT.COM 2000 exhibition in Bangalore. Singapore's participation was aimed at creating opportunities for its companies to build an in-depth knowledge of their Indian counterparts (*Business Times*, 2000).

In the same month, SingTel and Bharti Enterprises signed a US$650 million deal to build India's first private sector undersea cable (Saywell 2001a). The deal set up a 50–50 joint venture to build an 11,800 km fibre-optic cable called Aquanet to connect the Indian cities of Mumbai and Chennai with Singapore. In Singapore, the submarine cable would link with SingTel's other cable networks. France's Alcatel Submarine Networks and Japan's Fujitsu were selected to design, manufacture, install and commission the Singapore–Chennai leg of the submarine cable, which was expected to cost US $250 million.

When fully completed the proposed cable would be one of the world's largest submarine cables in terms of capacity, with a bandwidth of 8.4 terabits per second, capable of accommodating more than 100 million conversations simultaneously (*Business Times*, 25 October 2000). According to Lee Hsien Yang, 'this cable network will provide … the stimulus to drive the internet and e-commerce activities between India and Singapore and between these two countries and the rest of the Asia–Pacific region' (*Straits Times*, 25 October 2000).

SingTel's joint venture with Bharti in the undersea cable project was consistent with the former's overall strategy of acquiring or building stakes in foreign cable networks, as can be seen in Table 11.5. As of October 2000, SingTel had invested US$1.7 billion in undersea cables. The bandwidth demand in Asia is increasing by triple digits a year and the Indian market remains under-serviced. Accordingly, the new cable project is expected to pay for itself in a short time. Within India, Bharti has built a 10 000 km fibre optic network across 23 cities in Maharashtra and plans to build a 35 000 km network to connect 20–25 cities in Andhra Pradesh and Karnataka with Chennai.

Table 11.5 Cable projects in which SingTel has a stake

Name	Linking	Capacity (per sec)	Cost (US$ million)
C2C	Singapore, Japan, Taiwan, Hong Kong, Philippines, Malaysia, South Korea	7.5 terabits	2,000
Sea-Me-We3	Pacific Rim, South East Asia, Middle East, Europe	80 gigabit	1,400
APCN2	Singapore, Japan, China, Taiwan, Hong Kong, Europe	1.2 terabits	1,060
JV with Bharti	Singapore, Chennai, Mumbai	8.4 terabits	650

Source: Business Times, Singapore, 25 October 2000

The collaboration between SingTel and Bharti Enterprises illustrates the synergy and mutual benefits for both Singapore and India. For Indian companies such as Bharti, there is a pressing need for capital to finance their rapid growth. For example, to upgrade its existing mobile phone capacity, Bharti required US$318 million. In the event, it actually raised over US$1 billion in 2000 to cover this cost as well as to expand its network, consistent with its plan to become a pan-India mobile player. In that context, Singapore represents a highly suitable source of foreign capital for India, in view of the geographical location of the two countries relative to each other and the cultural ties that exist between them.

For SingTel and other Singaporean companies, the Indian market offers huge potential and capacity for growth. India's domestic and long-distance markets are estimated to be worth about US$6 billion in 2003. Yet potential for growth remains very high, because of India's large population and low telecommunications penetration rate, with only 2 per cent of the population having a fixed phone line and only 0.18 per cent owning a mobile phone. In addition, the average level of income and expenditure is likely to continue rising as a result of the ongoing economic reforms. Another factor which will contribute to growth in the demand for telecommunications services is the continuing rise of the 'new' economy, encompassing software development, call-centres and data-centres.

The growth expectations for India fit in well with SingTel's ambition to be the biggest telecom MNC in Southeast Asia. With a small population of only around 4 million, Singapore's own market is not big enough to generate the sort of economies of scale that SingTel's international competitors can achieve. Indeed, even within its own turf, SingTel has been forced to adjust to the entry of foreign players.

As a joint venture partner, Bharti Enterprises holds certain attractions for SingTel. It is a diversified group with stakes in the cellular, fixed-line, long-distance and internet interests. Importantly, Bharti has established a proven record of success in managing its businesses, forming partnerships and bidding for new licences. As a case in point, the group won almost half of the 17 new mobile licences granted after the July 2001 auction.

This is significant, as through its association with Bharti, SingTel has found the Indian market much more welcoming than Hong Kong or Malaysia. Indeed, SingTel had withdrawn from the Indian market in the mid-1990s, but its second attempt at entry has been far more successful, partly because of its choice of a partner. It has also been helpful that the new, reformist spirit has lessened some of the previous concerns about unclear regulatory policies (*Business Times*, 8 August 2000).

In India, besides its collaboration with Bharti, SingTel has a subsidiary called Singapore Telecommunications India with a presence in the four largest cities namely Bangalore, Chennai, Mumbai and Delhi. SingTel also promotes a range of telecommunications products and services in India and has started to explore with Bharti other investment opportunities overseas. As of March 2003, Dishnet DSL Limited (DDSL), India's leading ISP, became the first ISP to sign a contract with SingTel to procure bandwidth on the i2i cable. SingTel has agreed to supply DDSL with STM four international bandwidth capacity on this cable which connects to SingTel internet Exchange (STiX) in Singapore.

Besides SingTel, other Singapore firms are taking an active interest in the telecommunications sector in India. Two units from Singapore Technologies (ST), namely Singapore Technologies Telemedia (ST Telemedia) and Sembawang Corporation (SembCorp) were willing to pay more than S$47 million to acquire a 20 per cent stake in Modicorp, an Indian information and telecommunications services company. ST Telemedia and Modicorp have agreed to form joint ventures to establish internet data centres and offer long-distance telephony services in India. ST Telemedia would also nominate two of its senior managers to the Modicorp board and provide strong technology support (*Business Times*, 16 August 2000).

6. CONCLUSION

The synergy between Bharti Enterprises and SingTel is indicative of the potential benefits to both India and Singapore from cooperation as they meet the challenges of globalisation. India needs foreign direct investment (FDI) to expand vital sectors such as telecommunications and to develop its economy. Telecommunications form the backbone to the economy, as they provide essential support to many other sectors. The country's ability to attract FDI signifies a seal of approval from international investors.

For Singapore, India's telecommunications sector represents an opportunity as well as a gamble. While some regulatory risks remain, overall the market presents great opportunities for investment. These arise because of India's large and under-serviced population, and because of the considerable structural and policy changes that have taken place since 1991 and are still unfolding.

We close with some thoughts about the race to be among the major pan-Asian telecommunications companies. For SingTel it is clear that the race is not yet over. Nevertheless, its performance record while evolving into a S$5 billion telecommunications group with overseas investments in 20 countries compares reasonably well with many other companies, especially in view of the turbulence that has beset the telecommunications sector worldwide. Such illustrious companies as British Telecom, Deutsche Telekom, AT&T and Telstra have so far not been able to conquer the Asian market. Richard Li's PCCW was thought to have gained the advantage but stumbled when the Dot.Com bubble burst. Although in the past SingTel has often been berated by critics for being a laggard and slow to react, the company has managed to acquired stakes in significant mobile phone companies in India, Indonesia, Thailand, the Philippines and Australia. It navigated safely through the Asian financial crisis to emerge as an important player in the regional telecommunications market. Perhaps assessments of SingTel's performance by some of its severest critics can do with a little revision.

ABBREVIATIONS

BSNL	Bharat Shanchar Nigam Limited
CEO	chief executive officer
CII	Confederation of Indian Industries
COAI	Cellular Operators Association of India
DOT	Department of Telecommunications
DTS	Department of Telecommunications Services
EDB	Economic and Development Board
FDI	foreign direct investment
FTA	Free Trade Agreement
GDP	gross domestic product
ICT	information and communication technology
IDA	Infocom Development Authority
IE	international enterprises
IPO	initial public offering

ISP	internet service provider
IT	information technology
JV	joint venture
MCIT	Ministry of Communications and Information Technology
MNCs	multinational companies
MOU	memorandum of understanding
MTNL	Mahanagar Telephone Nigam Limited
NTP	National Telecommunications Policy
PCCW	Pacific Century Cyberworks
POTS	plain old telephone services
SingTel	Singapore Telecommunications Private Limited
ST	Singapore Technologies
StiX	SingTel internet Exchange
TDB	Trade and Development Board
TRAI	Telecommunications Regulatory Authority of India
VPN	virtual private network
VSNL	Videsh Shanchar Nigam Limited
WTO	World Trade Organization

REFERENCES

Asher, M.G. and R.S. Rajan (1994), *Singapore–India Economic Relations: Rationale, Trends, and Future Directions*, Centre for Advanced Studies, National University of Singapore, Pagesetters Services Pte Ltd, Singapore.

Banskota, N.P. (2002), *Building South Asian Economic Community. Lessons from North America and Western Europe*, Lewiston, NY: Edwin Mellen Press, Chapter 6.

Business Times (2000), Singapore, 16 August.

Business Times (2000), 'IDA leads contingent to India', Singapore, 25 October.

Business Times (2000), Singapore, 9 November.

Chen, Huey-tsyh and Eddie C.Y. Kuo (1995), 'Telecommunications and economic development in Singapore', *Telecommunications Policy*, September, pp. 240–44.

Chowdary, T.H. (1997), 'Telecommunications in India', in Eli M. Noam (ed.), *Telecommunications in Western Asia and the Middle East*, Oxford: Oxford University Press.

Divyanathan, D. (2002), 'Asian telcos are the world's most profitable', *Straits Times*, Singapore, 16 October.

Hansen, J. and Shameen, A. (2001), 'SingTel Blues, Singapore's top telco needs to win soon to become regional power', *Asiaweek*, 16 February.

Heng, T.M. and L. Low (1990), 'Towards greater competition in Singapore's telecommunications', *Telecommunications Policy*, August, pp. 303–14.

Huat, T.C. (1995), *Venturing Overseas. Singapore's External Wing*, Singapore: McGraw-Hill Book Co., Chapter 2.

Joo, T.K. (1993), 'The telecommunications situation in Singapore', in Marcel C. Werner (ed.), *Asia Pacific Connections. Telecommunications Assessment*, Amsterdam: IOS Press, Chapter 8.

Mani, S. (2000), 'Deregulation and reforms in India's telecommunications industry', in Mitsuhiro Kagami and Masatsugu Tsuji (eds), *Privatization, Deregulation and Economic Efficiency. A Comparative Analysis of Asia, Europe and the Americas*, Cheltenham, UK and Northampton, MA, USA: Edward Elgar, Chapter 8.

Mansell, R. and M. Jenkins (1992), 'Networks, industrial restructuring and policy: The Singapore example', *Technovation*, **12** (6), 397–406.

Mesher, G.M. and E.E. Zajac (1997), 'The political economy of telecommunications in Malaysia and Singapore: A structure–conduct–performance comparative analysis', *Information Economics and Policy*, **9**, 183–202.

Ministry of Communications and Information (2000), Singapore, Press Release, 27 January.

Saywell, T. (2001a), 'Singapore's new India play', *Far Eastern Economic Review*, 25 January.

Saywell, T. (2001b), 'Slow and steady at SingTel', *Far Eastern Economic Review*, 4 October.

Shari, M. (2001), 'SingTel is soaring beyond Singapore', *Business Week*, December.

Singh, J.P. (1999), *Leapfrogging Development. The Political Economy of Telecommunications Restructuring*, Albany: State University of New York Press, Chapter 5.

Singhal, R. (1997), 'Overall marketing activities in Indian telephone industries, Naini, Allahabad', in Devandra Thakur (ed.), *Economic Reforms and Industrialisation*, vol. 4, New Delhi: Deep and Deep Publications, Chapter 8.

Slater, J. (2002), 'Privatization: A second wind', *Far Eastern Economic Review*, 11 April.

Sridhar, V. (1998), 'Public sector enterprises. Challenges of change', *Frontline*, 10–23 January, **15** (1).

The Economic Times (2001), India, 23 February.

The Economic Times (2001), India, 24 July.

The Economic Times (2001), India, 16 August.

The Economic Times (2002), 'Basic telecoms services of poor quality: Trai Survey', India, 24 August.

The Hindu (2000), New Delhi India, 16 July.

Vittal, N. (1995), *India Incorporated. Reflections on the Indian Electronic Industry*, New Delhi: Vikas Publishing House Ltd, Chapter 43.

WTO (1997), 'Data on telecommunications markets covered by the WTO negotiations on basic telecommunications', (http://www.wto.org).

Yam, T.T.K. (2001), Opening address by Singapore Deputy Prime Minister and Minister For Defence, Techventure Asia, Hyatt Regency, Singapore, 18 September.

Yang, L.H. (2002), Breakfast talk by President and CEO of Singapore Telecommunications Limited, 'Growing out of Singapore – the SingTel experience', 20 February, Fullerton Hotel, Singapore.

Index